Ernst Schering Research Foundation Workshop 15
Computer Aided Drug Design in Industrial Research

Ernst Schering Research Foundation
Workshop 15

Computer Aided Drug Design in Industrial Research

E. C. Herrmann, R. Franke
Editors

With 51 Figures and 16 Tables

Springer-Verlag Berlin Heidelberg GmbH

ISBN 978-3-662-03143-8   ISBN 978-3-662-03141-4 (eBook)
DOI 10.1007/978-3-662-03141-4

CIP data applied for

© Springer-Verlag Berlin Heidelberg 1995
Originally published by Springer-Verlag Berlin Heidelberg New York in 1995
Softcover reprint of the hardcover 1st edition 1995

Typesetting: Data conversion by Springer-Verlag

21/3135–5 4 3 2 1 0 – Printed on acid-free paper

Preface

The Ernst Schering Research Foundation sponsored its 15th workshop in Berlin on October 19–21, 1994. Leading scientists from Europe and North America were invited to discuss computer-aided drug design in industrial research.

Computer-aided drug design is a very exciting field and an intellectual challenge, like playing chess. But these reasons are no longer sufficient to justify using this method in industry, if they ever were.

Fig. 1. The participants of the workshop

Therefore, when we, together with Prof. Hoyer, started to think about this workshop, our intentions quickly became clear.

We were not so much interested in the very latest developments of methods or in computer-aided drug design itself – enough conferences have dealt with these topics. However, we were very interested in the usefulness and limitations of computer-aided drug design in the industrial research process.

A lot has changed in the pharmaceutical industry recently. These changes are gaining momentum, so it is the right time to think about the role of computer-aided drug design in this changing environment. Just to mention two of these changes: First, most of us are faced with severe cost reduction programs and computer-aided drug design has to be judged by its possible impact on the bottom line; second, the significance of research is changing. Competition between pharmaceutical companies is much more pronounced nowadays. This competition is global and, even more importantly, it is *competition by innovation.* The title of an article in a research management journal highlights this quite clearly: "'Innovate or die' is the first rule of international industrial competition" (de Pury 1994).

The issues of cost reduction and competition by innovation formed the setting for our workshop, where we wanted to discuss the role of computer-aided drug design in the overall quest for a new drug, addressing topics such as how computer-aided drug design can be *integrated* in the overall research process to have the most impact, what computer-aided drug design can *contribute* and where the *pitfalls* are, and what factors *favor* or *prohibit* successful contributions from computer-aided drug design.

The goals set for this workshop were very ambitous, but we got much more than we ever expected. Indeed, this is reflected in these proceedings.

We gratefully acknowledge the contributions of the authors of the chapters in this book and the assistance provided by the Ernst Schering Research Foundation, in particular by Dr. Ursula-F. Habenicht.

E. C. Herrmann
R. Franke

Reference

de Pury D (1994) "Innovate or die" is the first rule of international industrial competition. Research Technology Management 37(5):9–11

Table of Contents

List of Contributors

C. Broger
Preclinical Pharmaceutical Research, Computational and Structural
Chemistry, F. Hoffmann-La Roche Ltd., 4002 Basel, Switzerland

D. Bur
Preclinical Pharmaceutical Research, Computational and Structural
Chemistry, F. Hoffmann-La Roche Ltd., 4002 Basel, Switzerland

D.M. Doran
Preclinical Pharmaceutical Research, Computational and Structural
Chemistry, F. Hoffmann-La Roche Ltd., 4002 Basel, Switzerland

E. Eckle
Institute of Physical Chemistry, Schering AG, Müllerstraße 178,
13342 Berlin, Germany

J. L. Fauchère
Institut de Recherches SERVIER, 11, rue des Moulineaux,
92150 Suresnes-Paris, France

R. Franke
Consulting in Drug Design GbR, 16352 Basdorf, Germany

P. R. Gerber
Preclinical Pharmaceutical Research, Computational and Structural
Chemistry, F. Hoffmann-La Roche Ltd., 4002 Basel, Switzerland

K. Gubernator
Preclinical Pharmaceutical Research, Computational and Structural
Chemistry, F. Hoffmann-La Roche Ltd., 4002 Basel, Switzerland

N. Heinrich
Institute of Physical Chemistry, Schering AG, Müllerstraße 178,
13342 Berlin, Germany

E. C. Herrmann
Institute of Physical Chemistry, Schering AG, Müllerstraße 178,
13342 Berlin, Germany

G.-A. Hoyer
Institute of Physical Chemistry, Schering AG, Müllerstraße 178,
13342 Berlin, Germany

R. M. Hyde
Department of Physical Sciences, Wellcome Research Laboratories,
Langley Court, Beckenham, Kent BR3 3BS, UK

H. Köppen
Medicinal Chemistry Department, Boehringer Ingelheim KG,
55216 Ingelheim, Germany

H. Kubinyi
Drug Design, ZHB/W, A 30, BASF AG, 67056 Ludwigshafen, Germany

Y. C. Martin
Pharmaceutical Products Division, Abbott Laboratories, Abbott Park,
IL 60064, USA

K. Müller
Preclinical Pharmaceutical Research, Computational and Structural
Chemistry, F. Hoffmann-La Roche Ltd., 4002 Basel, Switzerland

U. Norinder
Karo Bio AB, P.O.B. 4032, 14104 Huddinge, Sweden

T.M. Schaumann
Preclinical Pharmaceutical Research, Computational and Structural
Chemistry, F. Hoffmann-La Roche Ltd., 4002 Basel, Switzerland

J. P. Tollenaere
Theoretical Medical Chemistry, Janssen Research Foundation, 2340 Beerse,
Belgium

J. G. Topliss
Department of Medicinal Chemistry, College of Pharmacy,
The University of Michigan, Ann Arbor, MI 48109-1065, USA

H. P. Weber
Sandoz Pharma AG, Preclinical Research, 4002 Basel, Switzerland

1 Some Aspects of Computational Chemistry

G.-A. Hoyer

1.1 Introduction

The main goal of this workshop is not so much to present the latest scientific results but to talk about the advantages and disadvantages of computational chemistry in supporting research in pharmaceutical industry. I hope that we will come to the conclusion at the end of the workshop that computer-aided drug design is a necessary and important subdiscipline of chemistry and can generate many helpful ideas and insights in the field of structure–activity relationships.

In 1970 my former boss Dr. Röpke and I founded the group Theoretical Chemistry at Schering. We both had the strong feeling that theoretical chemistry was ready and able to support the bench chemists and biologists during their compound-finding activities. In spite of some negative statements by research managers we were able to convince the former chief executive director of research Dr. Raspé that it was worth spending money in this area.

Computational chemistry and rational drug design are great challenges to modern chemical research and compound finding. In the course of my introduction I will concentrate on the area of pharmaceutical compound finding. But everything I say is equally valid for agrochemical research.

Thirty years ago it was not yet routine practice to apply the methods of computational chemistry in the field of compound finding. Theoretical chemistry had already reached a high level, but experimental methods clearly predominated in pharmaceutical research. The search for new lead structures was carried out by means of so-called blind screening, the synthesis of vast numbers of chemical compounds and the testing of these compounds in various biological and pharmacological assays, and biological activities were optimized by systematic substituent variation. It was a procedure firmly based on the principle of trial and error.

It should, of course, be mentioned in this respect that great successes have been achieved in this way in the field of drug development. I need only remind you of the oral antidiabetics, the bactericidic sulfonamides, the large number of antibiotics, and the steroid hormones.

Research and development expenditure was high, though, and today it is enormous. The following figures make this clear. Nine years ago nearly 10 000 compounds had to be synthesized in order to produce one marketable active ingredient. Development lasted 8–10 years and cost approximately 100 million DM. Today the figures are as follows: 20 000–25 000 compounds synthesized, development lasting 10–12 years at a cost of 200–250 million DM. Thus, expenditure has increased considerably in the last 9 years and it will increase further if compound finding continues to be performed in the same way as before.

What possibilities are there of stopping or reversing this dangerous trend? Two routes are being taken by a number of pharmaceutical companies and divisions. First of all, the processes of life are being thoroughly investigated on a molecular biological basis in order to elicit new starting points for innovative active compound research and new biological methods such as bio- or gene technology are being employed to obtain new substances which could not otherwise be synthesized. The second approach is to apply the methods of computational chemistry in order to elucidate the relationships between chemical structure and biological activity and so to replace experiments by calculations.

Today, there is a third possibility, namely, to use the techniques of combinatorial chemistry, compound libraries, and high-capacity screening in order to figure out new lead structures in a very efficient way.

1.2 General Aspects of Computational Chemistry

The basis of all further considerations is finding out whether there is a relationship between the chemical structure of a compound and its biological activity and what form this relationship takes (Fig. 1). Conclusions as to the biological activity cannot be drawn directly from the chemical structure by mere analysis of the structural formula. But if one considers that each molecule possesses a shape in space, fills up a volume, and shows a distinct physicochemical behavior, and if one assumes, in addition, that the biological activity can only be observed when the active compound and the biological receptor interact with each other and form a stable complex for a while, then it is possible to make suggestions as to the structure–activity relationship. When I refer to physicochemical behavior, I mean the distribution of positive and negative charges, of polar and nonpolar zones and of hydrophilic and hydrophobic regions on the surface of the molecule, the capacity to form hydrogen bridges with the receptor, and the overall lipid- or

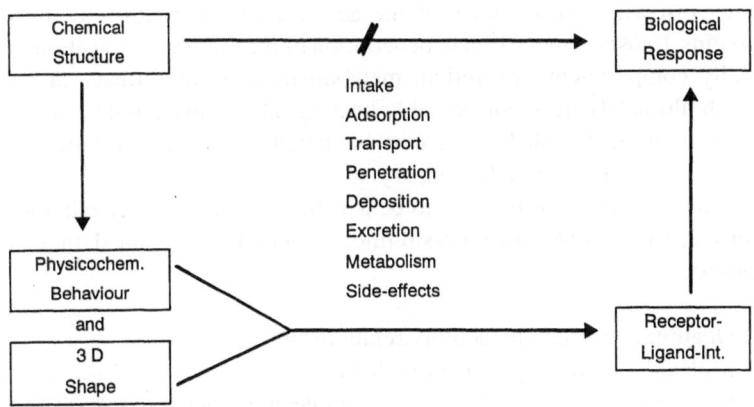

Fig. 1. Structure–activity relationships

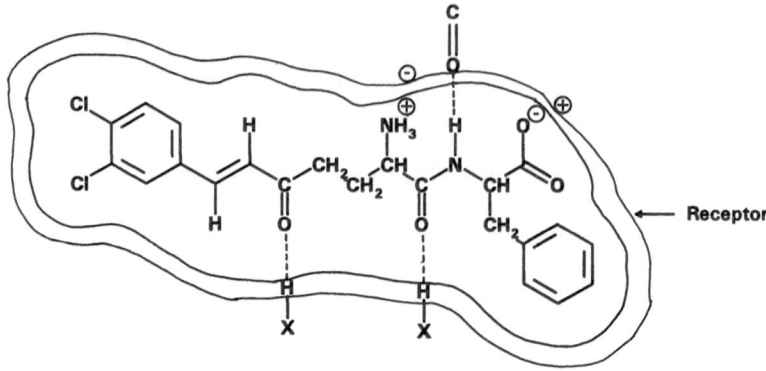

Fig. 2. Interaction between substrate and receptor

waterlike behavior. This can be demonstrated using a hypothetical active compound (Fig. 2). From a geometrical point of view the compound fits optimally into a depression on the surface of the receptor like a key in a lock; and its physicochemical behavior is complementary to that of the receptor. The decisive point is that the shape in space, the volume, and the physicochemical properties of molecules can be calculated with the methods of computational chemistry (Fig. 1), although it should not be forgotten that only direct interaction between the active compound and its receptor can be dealt with by way of calculation. Other effects, such as intake of the active compound, adsorption on proteins, transport in the blood, penetration of membranes, deposition in bodily compartments, excretion, metabolism, and side effects, all of which ultimately have considerable bearing on pharmacological activity, cannot be handled or cannot be handled satisfactorily by the methods of computational chemistry.

What methods can be used to deal with structure–activity relationships successfully? Three possibilities should be mentioned in this context.

1. Quantitative structure–activity relationships
2. Molecular modeling of small molecules
3. Macromolecular modeling of large molecules such as proteins and nucleic acids

My colleagues will present more details to illustrate my general statements in their lectures. But one important point should be emphasized: it is usually necessary to start the investigations with a known three-dimensional structure of the ligand–receptor complex. These data can be gained by experimental studies, such as protein nuclear magnetic resonance (NMR) or protein X-ray. On the other hand, studies of homologous proteins can give useful information, too. The theoretical chemist has to decide which technique should be used according to the knowledge available.

1.3 Application Software and Hardware

Let me now make some general remarks on the application software and the hardware.

As a rule we do not develop the software required in the field of computational chemistry ourselves, but buy it from reputable software companies. This keeps expenditures for long-term development within reasonable limits. One can roughly divide the mathematical tools which are necessary into three categories:

1. Empirical methods
2. Semiempirical methods
3. Ab initio methods

We used to join user groups in order to force the software vendor companies to improve and enlarge their software packages. In the past this path has been successful and this fact encourages us to stick to it in the future.

In order to apply the methods of computational chemistry and rational drug design successfully powerful computer systems are absolutely imperative. It is important that the theoretical calculations keep pace with the experimental investigations. It is no help at all if it takes weeks to perform the calculations and chemical synthesis and biological testing only take a matter of days. If this happens, then no bench chemist or biologist will be begging the computational chemist for his support.

The application of statistical programs to the investigation of quantitative structure–activity relationships only requires personal computers

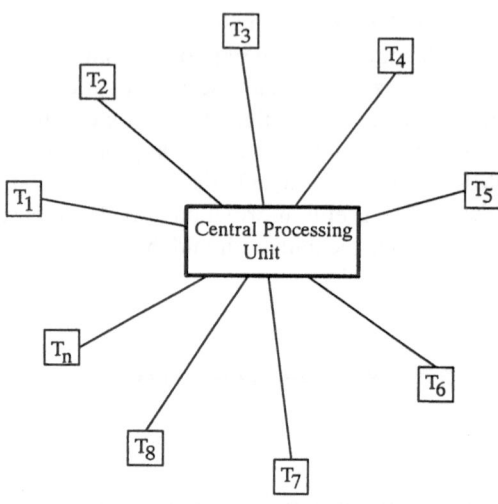

Fig. 3. Various terminals (*T*) linked to a powerful mainframe computer

or small workstations; for molecular modeling with force field or quantum chemical programs it is necessary to have powerful computers at one's disposal. The following tasks are especially time-consuming.

– All molecular dynamics calculations
– All quantum chemical ab initio calculations
– All molecular mechanics calculations of large molecules, such as proteins or polypeptides
– Molecular mechanics or semiempirical calculations on large series of compounds, i.e., series containing more than 50 compounds

In the past it was normal practice to install a powerful mainframe to which the various terminals were linked (Fig. 3). However, the central computer quickly became overloaded and performance got drastically worse with too many users and too many different tasks.

The next step was to install powerful workstations in the modeling groups (Fig. 4). These workstations are able to perform a large number of operations more or less on the spot. Today such workstations have a firm place throughout industry and the universities. Only the most

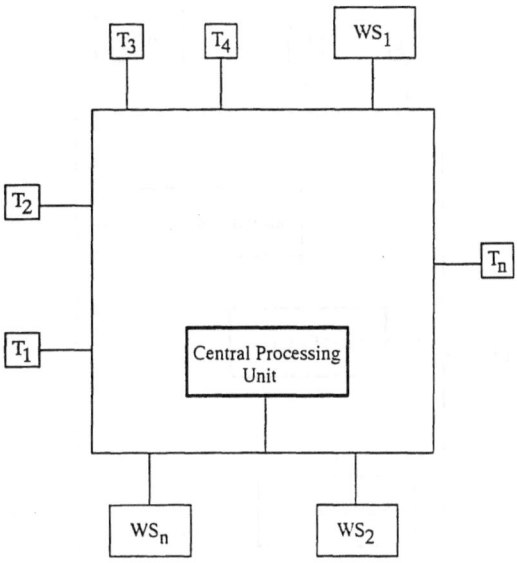

Fig. 4. Workstations (*WS*) in the modeling groups and terminals (*T*) connected by a net and linked to a central computer

time-consuming calculations are done on the mainframes which are also connected with the various elements of the computer configuration in a network in order to run the computers in an efficient way. However, the big problem is that nearly all mainframes are scalar computers and many theoretical calculations, especially the time-consuming jobs I have just mentioned, are performed too slowly on them.

Therefore one has to create a computer environment that allows most of the jobs to be run on extremely powerful workstations. The time-consuming tasks would be performed on special computers particularly suited to these jobs (Fig. 5). It would clearly be appropriate to use supercomputers for the time-consuming computational chemistry jobs and to use scalar computers only as file servers or to do simple operations or manage data banks. It goes without saying that all the elements of this computer environment would have to be standardized and compatible with each other in order to avoid communication problems.

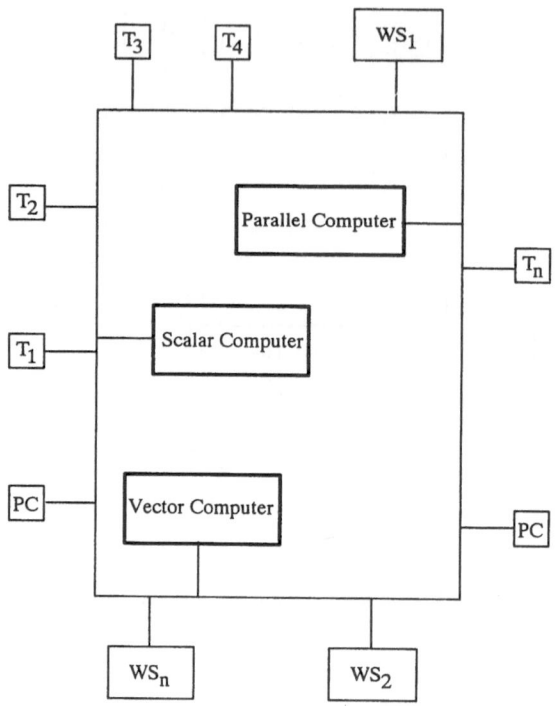

Fig. 5. Computer environment for specialized tasks. *WS*, workstation; *T*, terminal; *PC*, personal computer

1.4 Assessment of Rational Drug Design

During the last 10 years we have performed many investigations in the field of computer-aided drug design and we have gained a lot of experience, both positive and negative. At this point I would like to qualify my remarks in two ways.

1. In discussions with bench chemists, pharmacologists, and biologists I have often noticed that the colored pictures which you can find in many scientific journals simulate a reality which does not exist at all. All the approximations and shortcomings of the methods of

computational chemistry are incorporated in these pictures. The colored pictures are always only models.

2. Computational chemistry cannot predict drugs. There are too many other factors, which I mentioned at the beginning of my introduction and which decisively influence the success of drug development. These factors cannot be dealt with by the various methods of computational chemistry.

In spite of these two limitations I do think that rational drug design is helpful. It allows deeper insight into the mode of molecular biological action to be gained. It stimulates thinking and generates new ideas. It can replace experiment by calculation, rationalize compound finding, and shorten the path from the lead structure to the final active compound.

It should not be forgotten that computational chemistry can also be applied successfully in other areas of chemistry. We use these theoretical methods intensively in the calculation of physicochemical values and in the discussion of the chemical compound reactivity.

1.5 Outlook

In spite of all the shortcomings of the theoretical methods I think that the importance of computational chemistry and rational drug design in the field of compound finding will increase still further. Scientific development in this area is intense; this also applies for mathematical procedures, software, and hardware. Working in isolation will not bring success; positive results can only be achieved in close collaboration with other experts in the scientific community. Work in this field will not be cheap. The purchase and maintenance of hardware and software are expensive and this will remain so in the future. If one uses the theoretical methods critically, does not expect too much, and is always aware of the limits of computational chemistry, then the theoretical methods will provide a useful support for experimental research.

2 Computer-Aided Drug Design in Industrial Research – A Management Perspective

J. G. Topliss

2.1 Introduction

In recent years increasing attention has been devoted to the role and optimization of computer-aided drug design (CADD) in the drug discovery process. Two years ago a Workshop on the Computational Chemistry – Medicinal Chemistry Partnership for New Drug Discovery (Gund et al. 1992) was held and a follow-up round table discussion of the findings took place as part of the Medicinal Chemistry Division Program at the National ACS meeting in Washington, D.C., in the same year. A review on the establishment of CADD groups and factors affecting their contribution has also been published (Snyder 1991).

The purpose of my talk is to try to provide a general discourse on the management of CADD with the objective of making it a valued, productive, and resource-efficient contributor to drug discovery programs. Perhaps I should first explain my qualifications for addressing this subject so that you can understand the basis of my perspective. Although I am now in academia, most of my career has been in industrial research. In the 34 years that I spent in the pharmaceutical industry, from 1957-1991 with the Schering-Plough Corporation and Warner-Lambert/Parke-Davis, I have experienced a number of roles in new drug discovery research. These have ranged all the way from bench medicinal chemist and part-time QSAR practitioner to a succession of research management positions from Section Head to Vice President, Chemistry and membership in Pharmaceutical Research Management bodies. I initiated CADD activities at Schering-Plough and established a formal CADD program at Warner-Lambert/Parke-Davis and guided its development from a fragmentary operation in 1979 to a section comprising a director and 10 scientists in 1991. Thus, my perspective actually involves not only a management view of CADD but also reflects experience as a medicinal chemist using CADD and to some extent as an early CADD scientist.

2.2 Historical Background

CADD was born in the mid-1960s as quantitative structure–activity relationships (QSAR), following the publication of the landmark paper by Hansch and Fujita 1964, and began to be utilized by the pharmaceutical and agrochemical industries shortly thereafter. It was greeted by a mixture of enthusiasm and skepticism by the pharmaceutical medicinal chemistry community. In some companies significant efforts were mounted while in others it was virtually ignored. The value and potential of QSAR was greatly exaggerated in some quarters with simplistic visions of being a quick route to the next wonder drug and in other quarters it was dismissed as being without significant value. However, with further development and more experience in its use a much improved general understanding of its possibilities and limitations was arrived at and many successful applications were published, including some where it played a major role in the design of commercial compounds or compounds which reached a late stage of development (Fujita 1990). Recent extensions of what now is referred to as classical QSAR include CoMFA, a three-diemensional (3D) QSAR method (Cramer et al. 1988), and the use of neural networks (Aoyama et al. 1990). I always felt that the statistical/mathematical aspect of QSAR detracted from its popularity and use by medicinal chemists, hence, my focus on developing practical nonstatistical QSAR methods (Topliss 1972, 1977), but when computer-aided molecular modeling came on the scene in the late 1970s and early 1980s (Gund 1979; Humblet and Marshall 1980; Potenzone et al. 1977) it was quickly accepted because it fell more into the normal domain of molecular structural thinking to which organic/medicinal chemists were accustomed. This led to a strong, progressive growth in the use of CADD as computer power increased, improved software was developed, and more structural information on receptor targets became available from X-ray crystallographic and nuclear magnetic resonance (NMR) studies. In the last few years the scope of CADD has further increased and attention has turned to the more sophisticated exploitation of compound libraries through 3D pharmacophore matching and similarity/dissimilarity searching and methods for de novo drug design (Martin 1992; Humblet and Dunbar1993). Today, CADD is considered to be an essential element in any sizable drug discovery operation and all major pharmaceutical companies have a significant commitment to this technology. Over the past 30 years a considerable number of CADD methods have been developed covering a wide range of drug design problems, and now I think we have reached a much better understanding of what these methods can and cannot deliver.

2.3 What Should Expectations Be for a CADD Program?

It is important in a research organization for there to be a reasonable conformity of views on this question. If there are significant differences in expectations between CADD scientists, medicinal chemists, biologists, and management at various levels, serious difficulties are likely to arise. It is my view that the extremes of both over expectations and under expectations for CADD are largely behind us. This is a very stabilizing and healthy development because unrealistically high expectations, such as a quick route to a new drug candidate, leads to overemphasis and investment of resources in CADD followed by inevitable disillusionment and retrenchment, whereas unrealistically low expectations will not allow CADD to get the support and attention it deserves. What then should one reasonably expect from a CADD program? In a very general sense it must provide a net benefit to the drug discovery process commensurate with the resources invested. More specifically, CADD should be expected to provide new ideas to explore and hypotheses to test based on sound scientific rationales in the search for new active lead structures which would not otherwise be conceived or seriously considered. Thus, it should be able to add a whole new dimension to the idea pool for a project or program. CADD models should be reasonably predictive and able to classify possible synthetic targets according to degree of probability of success. Creative medicinal chemists can conceive of numerous plausible synthetic targets but often are able only to use ease of synthesis as a selection criteria. CADD can provide a complementary priority setting mechanism. It should also be expected that CADD can improve the efficiency of utilization of compound libraries for finding new interesting lead structures and additional compounds similar to existing lead structures.

2.4 The Organizational Position for CADD in Drug Discovery

Where CADD is positioned in a drug discovery organization is an important question relating to its effectiveness. The place CADD ends up in this respect can be a product of historical or special circumstances rather than a considered decision based on maximizing its contributions to the drug discovery effort. Although CADD needs to interface effectively with a number of entities the key link to pay attention to is that between CADD and medicinal chemistry since it is the medicinal chemists who will act or not act on ideas generated by CADD. In my experience this interaction is best fostered if these operations are both in the same organizational unit. CADD groups which are not situated in this way have often been ineffective in contributing to drug discovery efforts because of poor communication and interaction with the medicinal chemists. CADD also needs to have a cohesive link with structural chemistry (X-ray crystallography and NMR). The best arrangement, in my view, is to have all of these functions together in the chemistry department.

2.5 Composition of a CADD Group

For optimal operation a rather diverse set of skills needs to be assembled. Expertise should be present for both small and large molecule modeling, systems management, computer programming, theoretical and quantum chemistry, physical organic chemistry, organic chemistry, QSAR, statistics and compound database utilization. This list could be added to and is not presented in order of importance. Perhaps for various reasons some skill components must come from other organizational units in the company but here priority problems are often encountered and, in my experience, the more the necessary elements are present within the CADD group the greater is its overall effectiveness. Another option is to acquire missing components through outside consulting arrangements which can function as an external arm of the CADD group and bring in a high level of expertise not otherwise available. A further advantage of taking this route is flexibility of commitment depending on need. Individual CADD scientists typically possess some combination of skills and a group must be brought together which in aggregate embraces the necessary collection and balance of expertise to address a variety of drug design problems using the appropriate methodology. Apart from technical expertise, a CADD scientist in industry needs to have strong interactive and communication skills and the desire to work with scientists from other areas such as medicinal chemistry and structural chemistry in a team setting. Recruitment of CADD personnel with the desired qualifications is often a very difficult task, far more complex, say, than for synthetic medicinal chemists. There are no standard sources for CADD scientists as in the various branches of chemistry which have recognized departments in universities where the level and quality of the training is well understood. I have found that organic chemistry experience is very helpful for a CADD scientist in that it generates a better dialogue and mutual understanding with medicinal chemists, which is a critical factor in the overall success of a CADD program in a drug discovery effort. Also as we see ever greater and more complex applications of computational chemistry and molecular graphics to drug design problems, I have noted an increasing tendency to ignore the use of relatively simple physicochemical concepts which require minimal resources to implement and can be very productive. These tend to be dismissed as unsophisticated and old-fashioned – a serious mistake. It pays to have as part of the group someone with good physical organic chemistry skills who can provide this type of input.

2.6 CADD Group Size

How large should a CADD group be? Certainly over the years these have grown substantially as new CADD technologies have been developed and shown to be useful, and overall drug discovery efforts have increased. There are a number of factors to be considered in attempting to arrive at an answer from a management standpoint. CADD scientists point out that they are overwhelmed with projects either in progress or waiting to be worked on. On the other hand, medicinal chemists maintain that they have a large backlog of compounds worth synthesizing and no drug will be discovered until it has been synthesized. Both groups make a credible case for additional manpower but as always there is an overall resource constraint. Clearly, some sort of balance must be struck but what is the right balance? If we consider a research organization with 100 medicinal chemists (Ph.D.s and assistants/technicians) synthesizing compounds and a CADD group of 10 scientists there is a synthesis/CADD ratio of 10:1. Looking next at the relative costs of maintaining a synthetic chemist and a CADD scientist considering both salaries and chemical/supplies and equipment costs and including computer hardware and software licensing and maintenance costs, the CADD scientist is more expensive, perhaps by about 50% as a rough estimate (this will vary by company depending on size and level of hardware/software investment). Thus, the cost of the CADD operation is about 15% of that of the synthesis program and it could be judged to pay for itself if its input increased the efficiency of the medicinal chemists in synthesizing active compounds of interest by 15%. Similarly, if more conservative synthesis/CADD ratios of 15:1 and 20:1 are considered, the break-even efficiency increase required is 10% and 7.5%, respectively. Interestingly, many of the larger pharmaceutical research organizations appear to have synthesis/CADD ratios of between 10:1 and 20:1 and based on the foregoing analysis this seems to be well within the parameters of justifying the CADD investment. Indeed, it is possible that a larger relative CADD commitment would pay off, but to my knowledge this has not been adequately tested in large pharmaceutical companies. For a major therapeutic program effort covering several strategies and involving, say, 20 medicinal chemists, one CADD scientist equivalent might be assigned (this could be one full time or more than one each part time depending on the mix of expertise required). For smaller programs less than one CADD scientist equivalent would probably be committed. Also, it is likely that selected newer exploratory projects would benefit from some small level of CADD input. These probably would require a greater relative CADD contribution than for the major program. In addition, some allowance should be made for CADD new methods development. It can be seen that all of this would average out to somewhere in the CADD/synthesis ratio 1:10–20 range. Apart from this type of quantitative analysis, there are important intangible benefits to having a CADD program. What major pharmaceutical company would dare to be without one in this day and age? It would invite ridicule as a dinosaur organization, thus suffering image, morale, and recruiting problems in addition to being marked down by financial analysts for not possessing state of the art research technology! The critical mass factor which includes the ability to incorporate all of the necessary core skills must also be considered. Large organizations, particularly those with primarily centralized rather than scattered research operations, are in a favorable position here since they can accomplish this without unbalancing their investment in CADD relative to synthesis. Smaller organizations will have to have a proportionately greater investment in CADD or forego some aspects of it which would make it less efficient in addressing the full range of problems encountered in drug design. Looking at the resource effectiveness of the individual components of the CADD effort is a worthwhile thing to do. These can vary widely and an emphasis on those having a higher payoff can extract a bigger return for a given CADD group size. For example, modeling based on an X-ray crystallographic structure of a target–ligand complex is usually quite fruitful, whereas the same investment of resources in complex molecular dynamics simulations is likely to be less rewarding.

2.7 Integration of CADD in the Overall Research Process

The integration of CADD into the overall research process is greatly facilitated if, organizationally, CADD is in the same department as medicinal chemistry and structural chemistry. This should ensure that these activities are well coordinated, which is essential for maximum effectiveness. CADD should also have close links with protein chemistry and molecular biology so that when projects are undertaken which require the participation of all of these areas, the desired flow of activities will occur in a timely fashion. If the capabilities required to obtain the full range of structural information for CADD use are not available inside the organization then external collaborations need to be established for this purpose.

Another key aspect is for CADD to be an integral part of the drug discovery interdisciplinary team system and medicinal chemistry project groups, otherwise it will be isolated and will tend to regarded as an ivory tower, window dressing, or mere service organization.

2.8 Factors Favoring Successful CADD Contributions in Drug Discovery

A critical factor affecting the success of CADD in contributing to drug discovery is the support of top research management for a team approach involving all of the necessary players, including CADD. This support needs to be present through all levels of research management, down to the section level. If it is not, it is very easy for CADD to end up on the sidelines. CADD needs to be involved in the main interdisciplinary project team with the chemists and biologists and be a part of the medicinal chemistry working groups associated with the project team.

Both research management and the project team must understand the role of the CADD scientist and have a correct notion of what contributions are possible. If this is not done, unrealistic expectations of what CADD can deliver are likely to arise, with resulting problems. In this connection, it is especially important for the project team chair to properly appreciate both the potential and limitations of CADD.

The early involvement and integration of CADD into research projects, together with medicinal chemistry and structural chemistry, protein chemistry, and molecular biology is a key factor in promoting a successful contribution. CADD needs to be an integral part of the project from the outset. This is particularly important where it turns out to be possible to obtain structural information on a target protein. CADD can contribute best when the structure of the target itself or preferably the target bound to a ligand has been determined.

If involvement is delayed, CADD may not be able to participate effectively in a timely fashion. Another problem which arises with the late entry of CADD is that courses of action in the project become set and these may foreclose a useful role for CADD.

Successful CADD contributions are highly dependent on good frequent interactions between CADD and medicinal chemistry. This is a critical partnership which must be nurtured. I think that progressively over the years these two groups have improved their working relationships but it is necessary to be vigilant concerning this.

The extent to which CADD can contribute to drug discovery projects is also related to the range of methodological expertise present in, or available to, the group. The best CADD method to apply to a project will vary, depending on the nature of the problems posed, and so the broader the CADD capabilities are the greater will be the opportunity to make a contribution.

The location and type of facilities available for CADD can have a significant bearing on its effectiveness. Close proximity to medicinal chemistry and structural chemistry greatly improves interaction, as does the availability of conveniently located conference rooms and a nearby projection room to view structures in 3D.

It is important to make sure that CADD and medicinal chemistry have common objectives with respect to a project, otherwise there is little prospect for a successful outcome. A deep involvement and commitment to the project by the CADD scientist greatly improves the prospects for success since this not only produces a higher quality contribution but also focuses greater attention on the CADD approaches by the medicinal chemists. From the other side, if the CADD studies generate good practical ideas the medicinal chemists must be willing to follow up by synthesizing the necessary compounds to test the concept. Also, when joint plans are established these should be followed through to completion and side tracking should be resisted, e.g., a series design project which is of little use if not completed.

The quality and timely availability of biological and physicochemical data is a significant factor in the effectiveness of CADD. Precise biological data which are promptly available and the ability to obtain experimentally determined partition coefficient and pK_a data in a reasonable time frame when needed greatly improve the prospects for a useful outcome.

To help generate interest in, and understanding of, CADD by new medicinal chemists some kind of orientation program would be quite helpful and would promote the interaction process which is so important for making CADD productive.

Lastly, I should mention that a periodic, perhaps yearly, general discussion by CADD, structural chemistry, and medicinal chemistry on interface issues is very worthwhile.

2.9 Factors Hindering Successful CADD Contributions

To a large extent factors hindering successful CADD contributions are the converse of those discussed in the preceding section. Thus, an inadequate team approach, poor working relationship with medicinal chemistry, absence of agreement on project objectives, lack of structural knowledge of the target protein, and poor biological data all represent serious obstacles.

Failure to test CADD conclusions experimentally completely forestalls any opportunity for CADD to contribute successfully to a drug discovery project. A typical scenario here is that CADD staff is assigned to a project and develops ideas which are accepted by the medicinal chemists but are not worked on. Consequently, the project appears to have a CADD component but this does not factor in to the direction and results of the project and is in effect mere window dressing. The end result is a waste of CADD resources and a frustrated CADD staff. This type of situation is very damaging and it should be made clear to all involved that it is unacceptable.

An area where difficulties can be encountered is the selection of the appropriate method and/or software for the project at hand. This is not always straightforward and uncertainty and mistakes in method/software decisions can negatively impact CADD effectiveness.

Other hindrances are frequent hardware or software changes, which are disruptive and costly. Such changes need to be carefully analyzed and planned before implementation. Available software may not be adequate for the problem to which it is being applied and methodology limitations can also preclude a successful CADD outcome in a project.

A common problem is the CADD scientist taking on too many projects so that not enough attention can be given to each one. CADD scientists are very reluctant to turn down new projects, for understandable reasons – appearing to be uncooperative, lacking interest or missing future opportunities for example. However, getting overcommitted has insidious consequences which eventually come home to roost in the form of critical and disillusioned collaborators when progress on their projects is too slow. As a rule of thumb probably the most that should be attempted concurrently is one major project and one or two minor ones which may be in the formative or exploratory stage and have a long lead time. Other negative factors are lack of prioritization and uncertainties in degree of commitment to projects which CADD has been asked to work on. Also, time constraints which allow insufficient opportunity to bring projects to a fruitful stage are incompatible with successful outcomes.

2.10 Converting CADD Generated Ideas
into Synthesized Compounds

The impact of the follow-through or lack of it by medicinal chemists in synthesizing compounds suggested by CADD studies has already been mentioned, but the topic is important enough to return to in more detail. A number of situations can arise here which affect the final outcome. First the CADD hypothesis must be attractive and convincing and accepted by the medicinal chemists. This is where the communication skills of the CADD scientist are important. Good ideas are sometimes lost if not explained well. Assuming this hurdle is cleared, specific synthetic targets must be identified based on the CADD concept. The rule here should be that all suggestions, regardless of origin, be considered impartially and analyzed with respect to fitting the concept or testing the hypothesis and, equally important, as to synthetic accessibility. Some may be suggested directly by CADD scientists, especially those with an organic/medicinal chemistry background, although too many from this source might be counterproductive since the medicinal chemists would feel that their role was being usurped. Probably a larger number will come from the medicinal chemists and still others will arise jointly from the discussions by an iterative process. Trade-offs have to be decided between hypothesis fit and synthetic accessibility for it is seldom that the ideal compound from the hypothesis standpoint is easy to synthesize. Finally, a prioritized compound list has to be distilled out and synthetic assignments agreed upon. Rarely can all this be accomplished in one session – it usually requires several – and good follow-through here is essential to successful completion of this phase. All too easily the process can founder in delay and indecision.

Other factors enter the equation. The medicinal chemists will have other compounds they want to synthesize which are not CADD driven and the priority of this work in relation to the CADD-suggested targets must be established. Judgments have to be made concerning the relative attractiveness of proposed target compounds from all sources. However, care must be taken that good CADD-derived targets are not shut out in this process. They must receive a fair share of the synthesis effort. Here human nature comes into play. Medicinal chemists have their own ideas about synthesis targets and naturally they are often going to prefer to work on these. On the other hand, CADD scientists can be unrealistic in promoting their favorite target compounds which would be extremely difficult to synthesize. In resolving all of this much depends on the temperaments of the individual scientists, their working relationships, and the skill with which these interactions are managed. The approach which I think can minimize potential problems is to try to guide the process along collaborative and iterative lines so that the CADD scientists and the medicinal chemists both have a strong feeling of ownership and commitment to the target compounds agreed upon. The role of the CADD scientists will be primarily to generate and explain the design concept and describe the key molecular features required for the target compounds and the role of the medicinal chemists will be to translate this into specific, synthesizable compounds. To bring all of this off well requires watchful oversight at the appropriate management level but the rewards can be very substantial.

2.11 Setting Priorities

A good, well-run CADD group will soon have more business than it can handle. As already discussed, getting into an overcommitted position with regard to projects is a sure prescription for problems and must be avoided. What is the best way to handle this?

First, it is clear that the CADD group should continue to hold discussions with project team members concerning prospective new projects even though it is not in a position to initiate anything new at the time. To refuse such exploratory discussions would send the wrong message and inhibit future cooperation and interest on the part of the project team. The feasibility of a meaningful CADD contribution and its likely timing and scope can be assessed which provides valuable planning information for both parties.

In setting a priority for a project a number of factors need to be examined. Is it likely that CADD can make a useful contribution and, if so, at what resource cost? Would the timing of the CADD effort fit the needs of the project team? Do other groups such as structural chemistry and molecular biology need to contribute and, if so, what would the projected timing be for their work? What priority does the project team give the project among its various therapeutic area strategies? How does research management view priorities and timing across the various therapeutic area project teams? The answers to all of these questions must be considered in arriving at overall priority decisions, which are often not easy but need to be carefully arrived at if the maximum value is to be extracted from CADD activities.

2.12 Issues Encountered in Connection with CADD

Where to draw the line before making suggestions based on CADD studies for the medicinal chemists to act on can be a troublesome issue faced by CADD scientists. If this is done early, there is greater uncertainty with regard to the validity of the idea and, if done late, when the idea can be well supported, there is less possibility of influencing the project direction in a significant way. Thus, there is a fine line to tread here. Probably most CADD scientists err on the side of caution and are reluctant to put forward their proposals until they are quite sure of their ground. Preserving credibility is clearly very important but so is not missing the project boat.

To successfully navigate these shoals good communication is required between the CADD scientist and the medicinal chemists concerning the uncertainties of the proposed CADD model at various stages of the project and the risks of acting on any predictions based on the model at these different stages. An iterative approach with progressive model validation and refinement through compound synthesis and testing is an effective way to proceed and can productively begin quite early in the project cycle. Obviously, a thorough mutual understanding of the nature of the process and good judgment are the keys to a satisfactory outcome.

Another issue which arises is what constitutes success for CADD. In the early days expectations were unrealistic and success was defined as predicting the next wonder drug. I think that today some still view the scorecard on predictions as the primary success measure, but I believe most people involved now in the area of drug discovery understand that CADD is primarily a hypothesis and idea-generating mechanism and should be judged mainly on the value and utility of this type of contribution to the drug discovery process.

Inappropriate expectations from medicinal chemists and biologists arising from a poor understanding of CADD technology can be a troublesome issue and has been touched on earlier. Good communication and a suitable education should resolve these difficulties. A periodic seminar presentation by the CADD group can be very helpful in this respect.

Difficulties are sometimes encountered with regard to the credibility and acceptance of CADD ideas and proposals by the medicinal chemists. Sometimes the cause is a poor relationship between the groups through lack of cooperation and personality clashes or it may be due to inadequate communication and discussion of the CADD ideas. It might also be due to earlier disappointments with the outcome of past CADD proposals possibly compounded by unrealistic expectations or a perceived low quality of the CADD ideas presented. In any event, when this type of of situation occurs it requires prompt attention from management to ferret out the reasons and take the necessary steps to rectify the situation. This is much easier to do if both CADD and medicinal chemistry are in the same organizational domain.

Another issue which periodically surfaces relates to the time CADD staff have to spend " showcasing" CADD technology in demonstrations for various high-level corporate executives and other important individual visitors and touring groups. While this at times can be disruptive, I view this as a situation which has to be accepted and, in fact, this type of impressive presentation brings visibility to CADD, adds luster to the company's research image, and is a significant plus when large capital requisition requests for new equipment find their way up to the highest corporate levels for approval. Instead of complaints the focus should be on how to manage this activity efficiently so that disruptions and time spent are minimized. A good approach is to have various canned presentations at the ready to meet various requirements and if possible a separate area to do this in so that other work is not impacted. Also, it is necessary to use good judgment in filtering the requests which are made so that the load does not become too heavy.

There are a number of issues to be dealt with concerning software. The evaluation of new software offerings is obviously a necessary activity but can be very time consuming, so where does one draw the line on this? The selection of new software involves the evaluation of such questions as what new capabilities does it bring, compatibility and overlap with existing software, what does it cost, and will it be used enough to justify the cost. As we all know, budgets can quickly be eaten up in software expenditures, but yet, the right software addition can be enormously helpful. As mentioned earlier frequent changes in software are disruptive and software changes need to be carefully studied and planned. All of this represents a considerable challenge for CADD management.

2.13 Visibility and Recognition for CADD

It is necessary for research management to be watchful that CADD scientists receive the proper recognition and visibility with respect to their contribution to research projects. Since CADD scientists are not in a controlling position with regard to running projects, there can be a tendency to overlook their contributions, particularly if a strong, internally competitive environment exists in the research organization. Also, it is important that CADD be regarded as a true partner in research projects and not as just a service group. Again, it is the responsibility of management to ensure that the right attitudes are encouraged. Such considerations are important for the morale of the CADD group and, consequently, its performance.

When the project team makes presentations, if the CADD contribution has been significant the responsible CADD scientist should participate.

Along the same lines, project reports should contain a section describing CADD work on the project.

2.14 Benefits and Utility of CADD Contributions

The most direct type of CADD contribution to a research project results in the proposal of compounds for synthesis. As already mentioned, this is best done as an iterative process involving medicinal chemists which identifies target compounds meeting CADD model criteria and synthetic accessibility. Such a procedure promotes a "buy-in" by the medicinal chemists.

Another related type of CADD contribution is setting the priority of compounds for synthesis in the medicinal chemists list of compound ideas based on the degree of fit to the CADD model. This process may also result in the decision to exclude compounds for synthesis based on a very low success probability as judged by the CADD model. However, it is unwise to eliminate entirely from consideration a particular type of structural variant based on a CADD model which may turn out to be faulty. Instead, exploration of the structural variant should be restricted to one or two examples on a low priority basis and if these are inactive no others of that type would be made. This way it is possible to work with the probabilities of the situation without a blind, complete reliance on the CADD model.

A benefit of CADD is that it promotes a disciplined scientific approach to a problem by generating rational hypotheses based on available data which can be tested and modified as new data are obtained. In addition, it is often able to generate new ideas for a project which would not be arrived at otherwise.

CADD is also able to guide the selection of compounds from databases based on pharmacophore models, similarity or dissimilarity to reference structures, clustering techniques, or degree of fit to a defined receptor site.

2.15 Examples of CADD Contributions to Drug Discovery Programs

A detailed review of examples of CADD contributions to drug discovery programs is outside the scope of this lecture but, for the purposes of illustration, I do want to briefly mention a few examples which are taken from Parke-Davis Pharmaceutical Research Division projects.

The productive use of simple physicochemical ideas, requiring minimal resource investment, which I mentioned earlier in this lecture is illustrated by the use of the Hansch bioisostere concept utilizing substituent parameters (Hansch 1976) in the design of novel angiotensin (AT_1) antagonists. This concept provided the impetus for the novel replacement of a chloro substituent in a lead structure by a 1-pyrrolyl substituent (rarely, if ever, considered by medicinal chemists), on the basis of their similar hydrophobic and electronic parameter values, with retention of activity. Introduction of a 2-trifluoroacetyl substituent in the pyrrole ring of the resulting compound furnished CI-996 (Fig. 1), a potent, long-acting compound which was selected for preclinical development (Sircar et al. 1993).

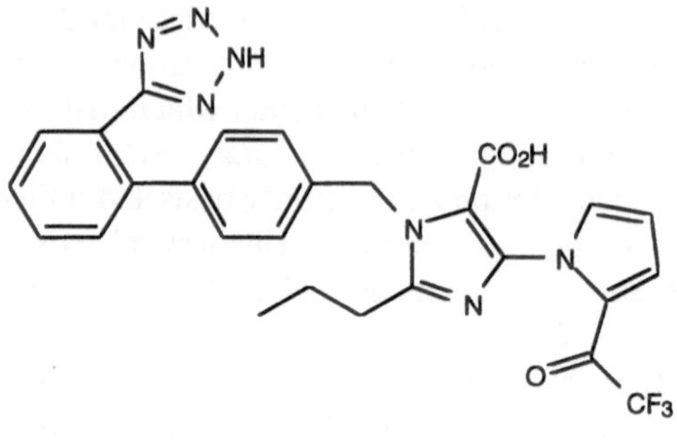

CI-996

Fig. 1. CI-996

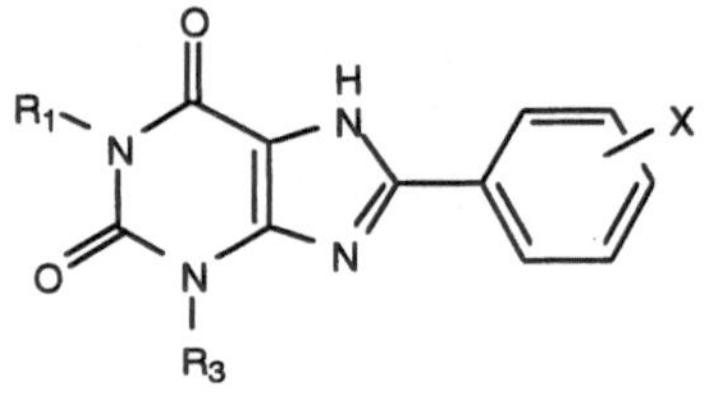

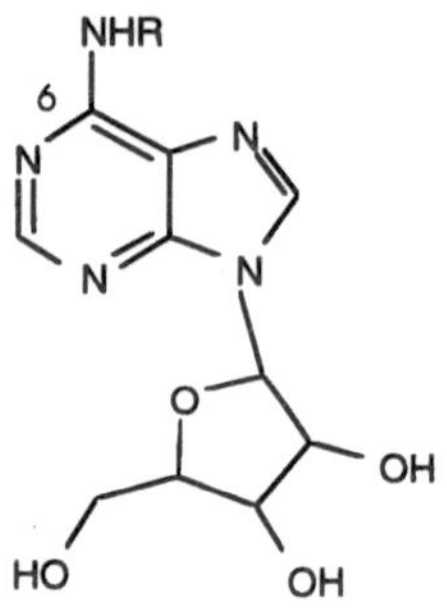

Fig. 2 *(left).* 8-Phenylxanthine adenosine antagonists

Fig. 3 *(right).* N^6-substituted adenosine analogs

The consideration of hydrophobicity as a global property is often important in drug design problems. An example of the relationship of hydrophobicity to tissue selectivity was provided in a recent study (Roth et al. 1991) on a series of 3-hydroxymethylglutaryl coenzyme A (HMG-CoA) reductase inhibitors which are an important class of drugs for lowering plasma total and low density lipoprotein (LDL) cholesterol in hypercholesterolemic patients. It had been proposed (Mahoney 1989) that the side effects of these drugs may be reduced by confining their action to the liver and that tissue selectivity is dependent on hydrophobicity with the more hydrophilic compounds showing higher liver selectivity. This hypothesis was tested by comparing a set of 15 potent, substructurally diverse inhibitors covering a broad range of calculated hydrophobicity (CLOGP) for their ability to inhibit sterol synthesis in rat liver, spleen and testis tissue. A linear relationship was found between tissue/liver ratios and hydrophobicity which showed that below a CLOGP of 2 the compounds are increasingly liver selective as they become more hydrophilic. This result supported the initial hypothesis and provided a sound basis for a synthetic strategy seeking more hydrophilic HMG-CoA reductase inhibitors.

The practical utility of QSAR studies is demonstrated by an application involving xanthines as adenosine antagonists (Hamilton et al. 1985). A QSAR analysis of a series of 56 8-phenylxanthine adenosine antagonists (Fig. 2) indicated that the most potent compounds had already been made, suggesting the termination of synthetic efforts to increase potency. Low aqueous solubility was a problem in the pharmacological evaluation of these compounds, and, since the QSAR analysis showed that potency was much more strongly affected by changes in *ortho* rather than *para* substitution in the 8-phenyl ring, an additional set of *para*-substituted compounds were synthesized, designed to increase aqueous solubility. These maintained high potency as predicted and aqueous solubility was dramatically increased.

Molecular modeling studies played a key role in the design of potent, highly selective A_{2a} adenosine agonists which were of interest as potential antipsychotic agents (Ortwine et al. 1990). A set of 15 potent N^6-substituted adenosine analogs (Fig. 3) were used to define a tolerated volume in the N^6-subregion of the A_{2a} receptor. Less potent analogs and a subseries of N^6-2,2-(diphenyl)ethyl derivatives were also examined. From these inputs an N^6-subregion receptor model was derived which was instrumental in the successful design of agonists with 3–5 nM potency and 12–30-fold selectivity for the A_{2a} receptor.

The design of tetrahydroisoquinolines as novel competitive *N*-methyl-d-aspartate (NMDA) antagonists was based on NMDA pharmacophore models established through molecular modeling studies (Ortwine et al. 1992). Using published NMDA ligands, the active analog approach was used in the generation of an agonist pharmacophore model. An independent modeling approach was used to derive a separate pharmacophore model for antagonists. These models are consistent with a single recognition site at the NMDA receptor that can accommodate both agonist and antagonist ligands and the antagonist model was used in the design of potent novel tetrahydroisoquinoline-competitive antagonists.

Molecular modeling based on analyses of ligand binding in crystal complexes is a powerful CADD method which has been applied in the design and evaluation of novel renin inhibitors (Lunney et al. 1993). In this work five renin inhibitors were cocrystallized with endothiapepsin, a fungal enzyme homologous to renin. The information gained from the X-ray crystal structures of the complexes provided size, shape, and polar requirement and restrictions in directing the docking of inhibitors in the active site of a human renin model and in the design of novel inhibitors. Potential inhibitors were designed and evaluated, assessing the polar interactions to guide docking and steric interactions with the enzyme cleft. The synthesis of certain analogs could be discouraged, based on blatant steric incompatibility with the enzyme and support for synthesizing compounds that were complementary to the binding site could be offered successfully on a qualitative basis. Also the extensive information gained from the analysis of the crystal structures was used in the de novo design of renin inhibitors.

Another example of molecular modeling used in conjunction with X-ray crystallographic data from the Parke-Davis experience relates to the elucidation of the binding mode of a novel nonpeptide HIV-1 protease inhibitor and its application in the design of related analogs (Lunney et al. 1994). An HIV-1 protease inhibitor, 4-hydroxy-3-(3-phe-noxypropyl)-2H-1-benzopyran-2-one (PD 99560; Fig. 4), was identified through the mass screening of an in-house compound library. A collaborative effort involving molecular modeling, X-ray crystallography, and chemical synthesis was then established to design and synthesize novel inhibitors based on the initial screening hit. The original prediction of the bound conformation of PD 99560 using molecular modeling was confirmed by one of the two binding modes observed in the X-ray structure of the inhibitor. As a result of this work the activity of PD 99560 was increased, albeit modestly because of limitations imposed by modes of binding. This lead was superseded as described in the following example.

In this closely related case (Prasad et al. 1994) mass screening of the Parke-Davis in-house compound library against HIV-1 protease uncovered 4-hydroxy-6-phenyl-3-phenylthio-2H-pyran-2-one (Fig. 5a) as

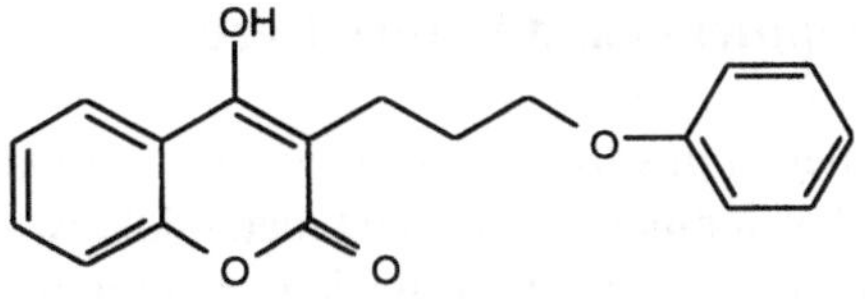

Fig. 4. PD 99560

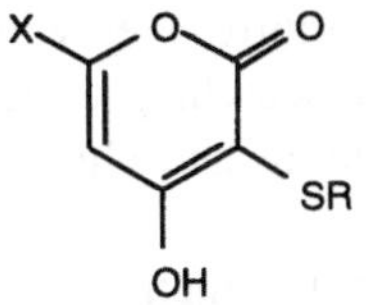

(a) X = R = C$_6$H$_5$

(b) X = C$_6$H$_4$(4-OCH$_2$CO$_2$H) ; R = CH$_2$CH$_2$C$_6$H$_5$

(c) X = C$_6$H$_5$; R = CH$_2$C$_6$H$_5$

Fig. 5a–c. HIV protease inhibitors

a low micromolar inhibitor. Active site binding interactions were proposed based on comparisons of the P_1-P_1' regions of known peptidomimetic–HIV protease complexes along with computer-based docking studies. Structure-activity and molecular modeling studies then led to a potent ($K_i = 51$ nM) inhibitor (Fig. 5b) where the 3-phenylthio substituent had been replaced by a 3-(2-phenethyl)thio group and a carboxymethoxy group had been introduced at the *para* position of the 6-phenyl substituent. This compound was noteworthy for its low molecular weight, absence of chiral centers, and synthetic accessibility in three steps. An X-ray crystallographic structure of the closely related 4-hydroxy-6-phenyl-3-benzylthio-2H-pyran-2-one (Fig. 5c) bound to HIV-1 protease was consistent with the initial binding hypothesis. In this complex, structural water-301, found in all peptide-based HIV protease inhibitor crystallographic structures, was replaced by the lactone moiety of the pyrone ring.

The examples I have discussed are representative of various types of CADD applications, involving different methodologies, in drug discovery projects. These demonstrate the value of having a range of methodological capabilities in a CADD group. They also provide some idea of the role CADD can play and the extent to which CADD input can contribute, in practice, to a drug discovery project.

2.16 Expected Developments and Future Needs

The current focus and interest in various CADD areas is a good indicator of where near-term developments in CADD methodology are most likely to take place. Future needs relate to inadequate currently available CADD methodology and technology which hamper or prevent the use of CADD in certain problems.

- The exploitation of databases will remain a high priority area with further improvements expected in software for the searching of 3D databases, including similarity/dissimilarity searching and clustering approaches.
- The current high level of activity in the development of methods for de novo ligand design can be expected to continue and to result in the future availability of progressively more useful and practical programs.
- Improved treatment of solvation and molecular dynamics is needed and developments along these lines can be expected.
- In the QSAR field we might look to further progress in the treatment of hydrogen bonding effects in relation to absorption/distribution phenomena and further extension of 3D QSAR methodology.
- In general, increased computer power should open new computational possibilities and enable greater model refinement.
- An important need is for new CADD methodology which can better address the various processes relating to the in vivo activity of compounds.
- There is also a need for user-friendly CADD software to be available in a suitably integrated desk top mode and some progress in this area is evident. Such desk top access will promote greater use of and familiarity with CADD methodology by medicinal chemists.

This list is not intended to be complete but rather illustrative of what might be expected in the future. There is good general agreement on this subject and a number of the items I have discussed have been mentioned in a recent CADD forum (Gund et al. 1992).

2.17 Judging the Effectiveness of CADD

Clearly, there is no single measure for judging the effectiveness of CADD in a drug discovery organization. Rather a number of different, primarily subjective, parameters need to be evaluated: First, is CADD well integrated with the drug discovery project teams and is there good interaction with the medicinal chemists? Next, the impact of CADD on compounds made in a project and their activity should be examined. How accurate were the activity predictions made as a result of CADD studies? Did CADD studies permit the paring down or prioritization of the medicinal chemist's list of compounds to synthesize in a productive manner? Was CADD able to provide interesting new structural ideas for the medicinal chemists to investigate? Were the CADD-inspired ideas practical targets for synthesis?

Authorship of papers and inventorship on patents is another measure of effective CADD contributions.

A more subjective assessment of CADD contributions is the ability to make a good case for an idea proposed through CADD such that it generates belief in the idea. Along similar lines the enthusiasm of CADD participants in projects is another factor which may relate to effectiveness. Finally, a quite subjective but nevertheless useful indicator is how the CADD contibution is viewed by the medicinal chemists on the project – with enthusiasm, positively, neutrally, or negatively?

An analysis of all of the above parameters collectively can provide, I believe, a useful measure of the overall effectiveness of the CADD contribution.

2.18 Improving CADD

There are a number of possible avenues for improving a CADD operation, depending on the prevailing situation. Many of the relevant factors have already been discussed. Better integration into the drug discovery teams and interaction with the medicinal chemists, if not already optimal, will be strong positives. Frequent face-to-face discussions with collaborators is very important and should only be supplemented by, not replaced by, other modes of communication. Changes in the physical layout or location of the CADD facilities may be helpful. The development of common standards for software would represent a significant improvement. Advances in technology and methodology will also result in improved capabilities and performance.

2.19 Should Medicinal Chemists Do Their Own CADD?

There are several different scenarios with respect to this question. The first is that CADD is a complex field best left in the domain of the specialist and medicinal chemists should confine themselves to synthesis. The second is that medicinal chemists should be strongly encouraged to undertake some CADD work, primarily modeling, along with their primary mission in synthesis. The third is to leave the choice to each individual medicinal chemist. Of these, I believe the third represents the best option.

In order for medicinal chemists to effectively undertake CADD they must have a strong interest in doing so. It is too complex for most medicinal chemists, and certainly in-depth studies should be carried out by CADD staff. Efficiency is a concern since with only intermittant use refamiliarization with the software may be necessary and new versions of software must be learned. A disadvantage is distraction from synthetic work and so a correct balance must be struck. Another difficulty may be the reluctance of medicinal chemistry section heads to divert effort from synthesis.

There are significant advantages to employing simpler modeling programs which are easier to learn and use. It allows the medicinal chemist to use these interactively as the project progresses so that ideas can be checked immediately and it improves the three-dimensionality of thinking. Also, this approach avoids the negative impact of a low priority assigned by the CADD group or delays in getting feedback. However, it is desirable to maintain some contact with the CADD staff concerning the project and at some stage, if the project becomes major, they need to be brought in to conduct more definitive studies.

Choice of the right modeling system is not without its problems. CADD scientists usually find fault with simplified modeling systems because they are not rigorous enough. Clearly a trade-off must be made between simplicity and rigor.

An advantage associated with medicinal chemists doing some modeling is increased flexibility of manpower utilization as project requirements fluctuate between modeling and synthesis. Additional plusses are improved understanding of the use and limitations of molecular modeling, the development of new ways of thinking about problems, and a strengthening of the interaction with the CADD staff.

2.20 Summary and Conclusions

Over a period of 30 years since its inception CADD has steadily advanced to become an essential component of drug discovery programs in pharmaceutical companies. Along with all the methodological and technological advances which have taken place over this time period we have also learned much about how to operate and manage a CADD unit so it delivers on its promise and expected productive contribution to the drug discovery process.

Key factors in this regard include a team approach to drug discovery which includes CADD, the support and understanding of research management, visibility and recognition, the organizational positioning of CADD, realistic expectations for CADD contributions, early involvement and integration of CADD into research projects, a broad range of methodological expertise in CADD, and frequent interactions between CADD and medicinal chemistry and common objectives in this partnership. It is important to look carefully at how to optimize the conversion of CADD-generated ideas into synthesized compounds, for if this is not done well, CADD is of limited value.

Some negative situations to avoid include not testing CADD conclusions experimentally, taking on too many projects and failure to set priorities.

Management needs to establish a good balance between deployment of drug discovery resources in CADD versus synthesis in particular but also in relation to other areas such as structural chemistry. Also the effectiveness of the CADD contribution to the overall drug discovery effort should be evaluated periodically and there are a number of parameters which can be employed in this regard. In addition, a periodic discussion between CADD and its partners in drug discovery on interface issues can be very helpful in identifying and correcting problems at an early stage and optimizing the interactive processes and productivity of CADD.

Not surprisingly, the opinions expressed here on many points are in accord with those reported as consensus viewpoints from a previous CADD Workshop (Gund et al. 1992).

In the future CADD has the potential to play a larger role in drug discovery than it does today. The degree to which this is realized will depend as much on how effectively CADD is managed as on new methodological and technological advances.

Acknowledgment. I wish to thank members of the CADD and medicinal chemistry staff from the Parke-Davis Pharmaceutical Research Division, Warner-Lambert Co., Ann Arbor, MI, USA for very helpful discussions in relation to the subject matter of this lecture.

References

Aoyama T, Suzuki H, Ichikawa J (1990) Neural networks applied to quantitative structure–activity relationship analysis. J Med Chem 33:2583–2590

Cramer RD, Patterson DE, Bunce JD (1988) 1. Effect of shape on binding of steroids to carrier proteins. J Am Chem Soc 110:5959–5967

Fujita T (1990) The extrathermodynamic approach to drug design. In: Ramsden CA (ed) Comprehensive medicinal chemistry, vol 4. Pergamon, New York, pp 540–544

Gund P (1979) Pharmacophoric pattern searching and receptor mapping. Annu Rept Med Chem 14:299–308

Gund P, Maggiora G, Snyder JP (1992) Workshop on molecular design strategies in new drug discovery: the computer-assisted drug discovery – medicinal chemistry partnership. Chemical design and automation news, pp 30–32; 204th American Cancer Society national meeting, Washington DC, 23–28 Aug. Division of medicinal chemistry abstracts nos 173 and 174

Hamilton HW, Ortwine DF, Worth DF, Badger EW, Bristol JA, Bruns RF, Haleen SJ, Steffen RP (1985) Synthesis of xanthines as adenosine antagonists, a practical quantitative structure–activity relationship application. J Med Chem 28:1071–1079

Hansch C (1976) On the structure of medicinal chemistry. J Med Chem 19:1–6

Hansch C, Fujita T (1964) ρ-σ-π analysis. A method for the correlation of biological activity and chemical structure. J Am Chem Soc 86:1616–1626

Humblet C, Dunbar JB Jr (1993) 3D database searching and docking strategies. Annu Rep Med Chem 28:275–284

Humblet C, Marshall GR (1980) Pharmacophore evaluation and receptor mapping. Annu Rep Med Chem 15:267–276

Lunney EA, Hamilton HW, Hodges JC, Kaltenbronn JS, Repine JT, Badasso M, Cooper JB, Dealwis C, Wallace BA, Lowther WT, Dunn BM, Humblet C (1993) Analyses of ligand binding in five endothiapepsin crystal complexes and their use in the design and evaluation of novel renin inhibitors. J Med Chem 36:3809–3820

Lunney EA, Hagen SE, Domagala JM, Humblet C, Kosinski J, Tait BD, Warmus JS, Wilson M, Ferguson D, Hupe D, Tummino PJ, Baldwin ET, Bhat TN, Liu B, Erickson JW (1994) A novel nonpeptide HIV-1 protease inhibitor: elucidation of the binding mode and its application in the design of related analogs. J Med Chem 37:2664–2677

Mahomey EM (1989) In abstracts of the Xth international symposium on drugs affecting lipid metabolism. 8–11 November 1989, Houston, p 103 (abstract 527)

Martin YC (1992) 3D database searching in drug design. J Med Chem 35:2147–2154

Ortwine DF, Bridges AJ, Humblet C, Trivedi BK (1990) Adenosine agonists. Characterization of the N^6–subregion of the adenosine A_2 receptor via molecular modeling techniques. In: Jacobson KA, Daly JW, Manganiello V (eds) Purines in cellular signalling. Springer, New York Berlin Heidelberg, pp 152–157

Ortwine DF, Malone TC, Bigge CF, Drummond JT, Humblet C, Johnson G, Pinter GW (1992) Generation of N–Methyl–D–aspartate agonist and competitive antagonist pharmacophore models. Design and synthesis of phosphonoalkyl-substituted tetrahydroisoquinloines as novel antagonists. J Med Chem 35:1345–1370

Potenzone R Jr, Cavicchi E, Weintraub HJR, Hopfinger AJ (1977) Molecular mechanics and the camseq processor. Comput Chem 1:187–194

Prasad JVNV, Para KS, Lunney EA, Ortwine DF, Dunbar JB Jr, Ferguson D, Tummino PJ, Hupe D, Tait BD, Domagala JM, Humblet C, Bhat TN, Liu B, Guerin DMA, Baldwin ET, Erickson JW, Sawyer TK (1994) A novel series of achiral, low molecular weight and potent HIV–protease inhibitors. J Am Chem Soc 116:6989–6990

Roth BD, Bocan TMA, Blankley CJ, Chuckolowski AW, Creger PL, Creswell MW, Ferguson E, Newton RS, O'Brien P, Picard JA, Roark WH, Sekerke CS, Sliskovic DR, Wilson MW (1991) Relationship between tissue selectivity and lipophilicty for inhibitors of HMG-CoA reductase. J Med Chem 34:463–466

Sircar I, Hodges JC, Quin J III, Bunker AM, Winters RT, Edmunds JJ, Kostlan CR, Conolly C, Kesten SJ, Hamby JM, Topliss JG, Keiser JA, Panek RL (1993) Nonpeptide angiotensin II receptor antagonists. 2. Design, synthesis,

and structure–activity relationships of 2-alkyl-4-(1H-pyrrol-1-yl)-1H imidazole derivatives: profile of 2-propyl-1-[[2'-(1H-tetrazol-5-yl)-[[1,1'-biphe-
nyl]-4-yl]-methyl]-4-[2-(trifluoroacetyl)-1H-pyrrol-1-yl]-1-H-imidazole-5
-carboxylic acid (CI-996). J Med Chem 36:2253–2265

Snyder JP (1991) Computer-assisted drug design, part 1. Conditions in the
1980's. Med Res Rev 11:641–662

Topliss JG (1972) Utilization of operational schemes for analog synthesis in
drug design. J Med Chem 15:1006–1011

Topliss JG (1977) A manual method for applying the Hansch approach to drug
design. J Med Chem 20:463–469

3 Reflections on Collaborations of a Computational Chemist with Medicinal Chemists and Other Scientists

Y. Connolly Martin

3.1 Overview

My responsibility is to lead a group that applies molecular modeling and quantitative structure–activity relationships (QSAR) concepts to make the process of drug discovery more efficient. Our emphasis is on applying known methods to problems of immediate interest to Abbott. However, since the methods to accomplish this are not perfected, we must continually evaluate and refine old methods and develop new ones. In this way we will be more able to answer the questions of key concern to the projects. This report will illustrate how interaction with other scientists has affected the accomplishments we have made.

It is my thesis that computational chemists have the largest impact on the success of projects if they are true collaborators with the corresponding synthetic chemists. The synergism between the two groups is enhanced when they become true collaborators by sharing in the responsibility for setting and accomplishing mutual goals. In my experience, good collaborations provide inspiration and stimulation to their participants and lead to improvements in computer-assisted molecular design (CAMD) methodologies as well as more potent molecules. Clearly, this synergism cannot be accomplished unless there is good communication between the experimental and computational scientists.

What do I mean by collaboration? Collaborators work together on a single project sharing responsibility for planning and executing the research, for defining what is success, and for its success or failure. Collaborators must communicate freely and honestly with each other, criticizing and questioning every aspect of the work while still recognizing the other's right to do the same and to make their own decisions. Collaborations are characterized by intense and frequent communication and mutual education of the participants. To collaborate includes the option to conclude or reduce the intensity of the collaboration if other research appears to be a better use of one's time.

In contrast, colleagues work in the same place but do not share responsibility for setting and accomplishing common goals. Colleagues might work on the same project – for example, one might synthesize compounds for testing, another test them, and a third develop QSAR. However, they remain colleagues and not collaborators if they share only results and not responsibility for developing common goals and strategies. While a team of colleagues can be successful, their chance for success is enhanced when they work more closely together.

In this report I will describe several interactions that led to important scientific results. They were chosen because they illustrate important points about the dynamics of collaborations and the impact of the synergism of people attacking the same problem from different points of view. Two will be discussed in detail; others more briefly. I will not consider in detail those relationships that were not successful because it often is hard to pinpoint the reasons for this. However, I will provide some personal generalizations concerning the atmosphere that fosters collaboration.

Since this is a personal account, other's experiences may differ and other's recollection of the same events may also differ.

3.2 Collaboration with a Synthetic Chemist Leading to a Novel Series of Potent D1 Dopamine Agonists (1988–1991)

3.2.1 The Original Problem and Its Solution

Soon after Abbott started a project to define a D1 selective dopamine agonist, a synthetic chemist on the project approached me for modeling help. He pointed out that the pendant phenyl group of SKF 38393 (compound II, Table 1) adds more than 100 times to its affinity for the D1 receptor while slightly decreasing affinity for the D2 receptor (Table 1). His question was, where is the pendant phenyl group with respect to the presumed pharmacophoric N and OH?

Molecular modeling by distance geometry and MMP2 optimization revealed that SKF 38393 assumes two low-energy conformations plus

Table 1. D1 and D2 receptor affinity

Compound no.	Name	R	D1 pK$_i$	D2 pK$_i$
I	A-54756	H	5.0	5.6
II	SKF-38393	C$_6$H$_5$	7.2	5.2

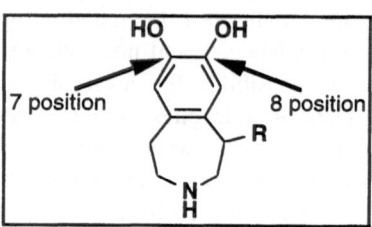

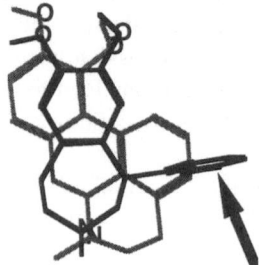

Fig. 1. The superposition of the equatorial conformer of I over the reference compound III. The *arrow* indicates the position of the pendant phenyl group

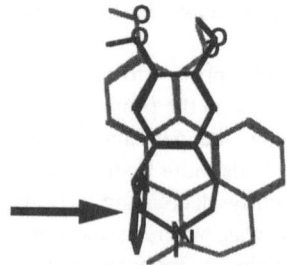

Fig. 2. The superposition of the axial conformer of I over the reference compound III. The *arrow* indicates the position of the pendant phenyl group

III

other conformations that resemble one or the other of these. Both have an N–O distance that matches the D2 pharmacophore. The two low-energy conformations differ in that, in one, the phenyl group is axial and, in the other, it is equatorial. In the equatorial conformation the 8-OH group is used for the pharmacophore, with the result that the phenyl at C1 superimposes somewhat with the extra phenyl of apomorphine (III) (Fig. 1). However, in the axial conformation the 7-OH group is used and the phenyl is in a new region of space (Fig. 2). Which is the correct superposition or location of the pendant phenyl group?

The structure–activity relationships reported in the literature show that replacing the 7-OH group with a Cl leads to a compound that binds even more strongly to the receptor even though it is an antagonist. This result would suggest that the equatorial conformer is the bioactive one. However, solution nuclear magnetic resonance (NMR) of the isolated molecule showed that the axial conformer is preferred by at least three-fold. My collaborator and I agreed that we must design and synthesize molecules to specifically answer the question.

My collaborator suggested that he make pairs of compounds. One member of each pair would place a phenyl in the space of the phenyl in either the axial or equatorial conformation of II. The other member would be devoid of the phenyl. Thus, by comparing members of these pairs we would see in what region in space a phenyl group adds to affinity.

The two of us worked closely to design the compounds. This involved almost daily discussions until we settled on the set of molecules that both answered the structural question and could be synthesized. It took about a year to make all the compounds since each pair requires a different synthetic pathway. Although none of the designed compounds was very potent, the comparison of the pairs of molecules indicated that in the bioactive conformation of II the pendant phenyl group is equatorial (Martin et al. 1991; Martin et al. 1993b).

3.2.2 Concurrent Work Related to Dopaminergic Agents

The design of novel compounds is of no use if one cannot predict their potency rather accurately. Hence, in the modeling group we were also actively seeking a three-dimensional (3D) QSAR method for this purpose. Although our own strategies were not successful, we had learned of Comparative Molecular Field Analysis (CoMFA) (Cramer et al. 1988) before it was published and had been applying it to Abbott's α-2 adrenergic agonists and D2 dopaminergic agonists reported in the literature before the D1 effort started and concurrent with it (Martin et al. 1993b).

3.2.3 The Big Payoff

Once we knew the D1 pharmacophore, my collaborator and I were eager to discover a new series of molecules. However, I was disappointed that the design proved to be extremely hard – it seemed impossible to use molecular graphics to design molecules that position the required basic nitrogen, catecholic OH, and pendant phenyl in the correct geometric relationship. This frustrating experience convinced me that we must write a computer program that searches databases of 3D structures to find those that meet geometric criteria (Van Drie et al. 1989). The objective would be to subsequently modify the database hits to include the required N and OH groups (Martin and Van Drie 1993) – this will be described in the next section.

The techniques of 3D searching and computer de novo design were invented because of our collaborations with a synthetic chemist – they have proved to be extremely valuable to the field (Borman 1989; Martin 1992). If our modeling group had just been "developing methods," we probably would have merely improved existing ones rather than recognize that a whole new approach was needed. If our role had been as a "support group" (i.e., not part of the inner circle that set the strategy for the project), we would not have identified with the need to design new

Fig. 3. The bioactive conformation of II and V

molecules and so also would not have taken the risk to develop the new approach. Similarly, without the strong likelihood that designed molecules would be synthesized, the chances of developing a new methodology would also have been lessened.

Concurrent with our modeling and synthesis efforts, the biochemists on the team had identified a nonselective dopaminergic agonist (IV) by testing catecholamines made for our adrenergic agents project. The project leader had been watching us enthusiastically design weakly active and inactive compounds. However, he trusted what we were doing to the extent that he asked me where to put the phenyl on that compound to make it nonselective. From 3D modeling there was only one choice (V), Fig. 3. The resulting compound was equal in affinity to II, and it is a full, rather than partial, agonist (Martin et al. 1991).

Our early prototype 3D searching program identified a number of Abbott catecholamines that met the dopamine pharmacophore require-

IV. R=H
V. R=C₆H₅

VI. R=H
VII. R=C₆H₅

VIII. R=C₆H₅

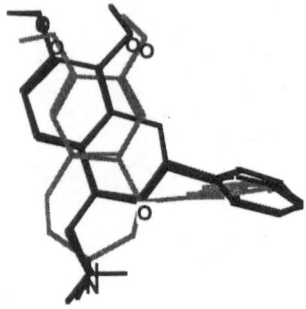

Fig. 4. The bioactive conformation of II, VII, and VIII

ments but had not been tested for dopaminergic activity. Several of these compounds bound to the D1 receptor and the synthetic chemists added the phenyl group to one of these (VI and VII) with the expected increase in affinity. Since this type of compound was relatively easy to synthesize, an extensive SAR was developed. This led to the isochroman series (VIII), the most potent and selective D1 agonists known (Schoenleber et al. 1990; DeNinno et al. 1991). VIII is more than 100 times more potent than VII. VII and VIII are shown in Fig. 4.

3.2.4 Follow-up

Although the emphasis of the synthesis effort was put on the isochroman series, the project wanted to have another totally different series as a back-up. I used Aladdin de novo design to suggest such compounds (Martin 1990). Then the whole team of synthetic chemists considered and refined these suggestions to identify compounds that they thought could be synthesized. They also sometimes used Dreiding models to design compounds. The modeling group then examined the conformational possibilities of these proposed compounds and forecast their affinity and selectivity with our concurrently derived CoMFA model for D1, D2, and α-2 pK_i (Martin et al. 1993; Martin 1994).

In total we used CoMFA to forecast the affinity of 201 proposed compounds. For 55, or only 27% of the total considered, the pK_is were

Table 2. Results of biological testing

| Forecast pK$_i$ | No. modeled | No. made | Observed pK$_i$ | | |
			< 6	6–7	> 7
Below 6.5	146	9	6	3	0
Above 6.5	55	10	1	4	5

forecast to be in the interesting range of 6.5. Nineteen of the 201 compounds were synthesized: nine with forecast potency and 10 with forecast potency 6.5. The nine compounds with low forecast affinity that were synthesized were generally made to test the model or to gain other important structure-activity information. Hence, the synthetic chemists believed the model enough to use it to set priorities for synthesis.

The result of the biological testing is shown in Table 2 (Martin et al. 1993; Martin 1994).

The model successfully separated the potentially potent from inactive compounds – if the objective is pK$_i$ 7, then half of the compounds forecast to be potent in fact met this criterion, whereas none of those forecast to be less interesting met it. The only serious error was one compound that was forecast to have a pK$_i$ of 7.4 but was observed not to bind even at 10^{-3} M. We thus had a model with predictive ability as well as a computer program that designed molecules.

3.3 A Collaboration with a Computer Specialist that Led to the Development of 3D Searching and De Novo Drug Design (1986–1990)

3.3.1 The Original Problem and Its Solution

After a few years working in molecular modeling, I realized the need to organize our 3D structures. In particular, we needed to have them in a chemical information database so that all conformers of a particular molecule could be retrieved by a name, these conformers could be used to construct conformers of analogues, we could identify the selected bioactive conformer of a molecule so that others could retrieve it easily, we could keep track of other associated files of a molecule, and biological activity or other properties of the whole molecule could also be searched.

I did not use the software from Molecular Design, Ltd. (MDL) because one could not integrate its chemical information capabilities into another program. I selected Dr. David Weininger as a collaborator because he had successfully programmed CLOGP for the MedChem Project at Pomona College and had developed a chemical information system to store octanol-water logPs and pK_as (MedChem Software Manual Release 3.54, 1989). Moreover, he wanted to use the computer to solve chemists' problems and I felt that he had the creativity and drive to develop novel solutions to problems.

Dave agreed to collaborate with us because he thought it was an interesting problem. On his first visit, he looked over the situation and wrote the prototype of a program that converts 3D coordinates into a SMILES/TDT that can then be loaded into the MedChem database system THOR. Abbott's only contribution was information on the format of the input files, the van der Waals radii of atoms, and enthusiasm for testing the program.

Dave insisted, to my frustration, that we reorder the coordinates and store them in the order of the atom in the unique SMILES string. I objected because the result would be that the atoms in my molecules would have different names than I was used to, but decided that probably Dave had reasons that I did not understand to insist on this. We continued to perfect the program but were quickly pleased with its ability to organize our data. We achieved our goal in a few months of low-level collaboration (Martin et al. 1988).

3.3.2 Concurrent Collaborations Between the Two Groups

Dave saw what we and others were doing with computational chemistry and decided to write a program, GENIE, that detects user-specified 2D substructures in molecules. One uses the associated GENIE Control Language (GCL) to program the features to identify and what to do if they are found. I eagerly tried GCL and suggested additional features for the language.

We used GCL to write a simple expert system to build analogs from a previously selected bioactive conformation of a lead molecule (Martin et al. 1986). However, the GCL programming took more time than doing ordinary molecular modeling so we did not pursue this idea. This phase was just exploration of tools without knowing where they would lead.

We ultimately integrated the MedChem software into our molecular graphics program (Martin et al. 1992). This allowed us to retrieve coordinates directly from the chemical information database to the modeling system. With the supplied algorithms to generate a unique SMILES for any structure we were able to name the atoms in a molecule the same way, no matter what their source. With the GCL substructure perception language we were able to write molecular graphics macros that could be used on molecules of disparate structures – for example, to color all phenolic hydrogen-bond donors blue, we would give the commands

COLOR * atom/[O,S,N;H] blue
COLOR * atom/H-[O,S,N;H] blue

This would process all molecules (*) currently in memory and change their color to blue if they are O's, S's, and N's that bear a hydrogen or if they are the corresponding attached hydrogen. More complex definitions could be accessed by reading in a file of definitions. We used this facility for many years and insisted that it be available when we switched to UNIX workstations.

3.3.3 The Big Payoff

As mentioned above, once the D1 project team knew the geometric relationship between the phenyl group, the basic nitrogen atom, and the catecholic oxygen in the D1 pharmacophore, we wanted to design new molecules. As I tried to do this at the graphics screen, I became more and more frustrated since I did not seem to be able to design molecules with the correct geometry.

Reflecting on this experience, I realized that the molecules that I had been building were selected from searching my mental database of 3D structures. Why not write a computer program to do this – to find templates to position the pharmacophoric groups? Since we were already storing 3D structures of molecules in a chemical information database and GENIE could recognize substructural features of molecules and which atoms were associated with these substructures, such a program should be possible. So, I brought in an outside collaborator, John van Drie, who designed and programmed the 3D searching program Aladdin (Van Drie et al. 1989).

As detailed above, a very early version of Aladdin identified approximately 30 molecules that match the dopamine pharmacophore but had not been tested for dopaminergic activity. So many of these compounds were active that the biochemist walked, rather than phoned, my office to request "more like the compounds you just suggested."

It was not easy to meet this request. Since we had already searched all 3D structures we had modeled and had submitted all the hits for which we could find sample, we tried to find samples of those "hits" for which the supply was supposedly exhausted. We found a few in pharmacologists' labs, but not many.

Another way to use the 3D pharmacophore to suggest more samples for screening would be to search more 3D structures. Only later did CONCORD (Pearlman et al. 1988) become commercially available – Abbott was one of the first companies to purchase it. However, this was too late for the D1 project. This strategy has been successful in a number of instances (Martin 1992).

The final alternative was to design new compounds. Recall that the original objective of Aladdin was to help us design new molecules. John van Drie invented an additional routine, MODSMI, to redesign molecules (Martin and Van Drie 1993). With the MODSMI language one can remove, replace, or add atoms, groups, or bonds to a 2D structure. We programmed it to convert the database molecules into those with the dopamine pharmacophore. Since MODSMI operates on 2D structures, we typically generate the coordinates of the redesigned molecules with CONCORD.

I applied the Aladdin/MODSMI strategy to designing potential dopaminergics (Martin 1990). Searches of thousands of 3D structures led to the design of 384 potential dopaminergics. These fall into 64 unique families (Martin and DeLazzer, in press). In contrast, the medicinal chemistry literature includes only fifteen families of compounds. So, computer de novo design does suggest new compounds.

Our collaborating synthetic chemists examined the designed structures and responded with those, or analogs, that they thought might be synthesizable. The results are listed above.

3.3.4 Spin-offs

We have used Aladdin for many years, adding functions to automate many of the tasks that are typically done manually with molecular graphics. Our most recent enhancement to Aladdin was associated with our automated pharmacophore mapping program DISCO (Martin et al. 1993a). We use Aladdin to identify atoms that can potentially participate in hydrogen bonds or charge interactions or to identify hydrophobic groups. The recognition is written in GCL: Aladdin then uses the coordinates of these atoms to calculate the ideal location of potential complementary points of the bio-macromolecule.

Dave went on to form Daylight Chemical Information Systems, a highly successful new software company. The rewritten version of the software, an object-oriented toolkit style, is an ideal tool for the Aladdin sort of application program. Presumably the collaboration provided Dave with some good ideas, too.

3.4 Examples of Less Dramatic, but Interesting Contributions I Made to Drug Discovery Projects

3.4.1 Calculating the LogP of Terasosin (1975)

At the request of a senior medicinal chemist, we calculated the octanol-water partition coefficients of α-1 adrenergic antagonists and noted a weak positive correlation with logP and antihypertensive activity (these were the days before receptor binding was an acceptable test). However, there was a large difference in intercept between different series.

Reasoning that the deficiency in the competitor's compound prazosine (IX) was its poor water solubility, the synthetic chemist asked me if logP calculations supported his hypothesis that replacing the furan with tetrahydrofuran (X) would increase water solubility. The calculations confirmed his theory; he made the compound, and it is now one of Abbott's best sellers.

Would he have made the compound without my input? Probably.

IX	C_4H_3O
X	C_4H_7O

3.4.2 Diuretics (1978)

A number of years ago Abbott was involved in a project to discover a uricoscuric diuretic. The compound reported in the literature was tielinic acid (XI). Rather early in the project the medicinal chemists discovered that they could replace the thiophene with a *para*-phenol (XII). However, the compound was not as potent as they wished and was poorly soluble. I was involved in selecting substituents for the phenolic ring so that we could derive a QSAR.

XI.	
XII.	$R_1 = R_2 = H$
XIII.	$R_1 = R_2 = CH_2N(Me)_2$
XIV.	$R_1 = CH_2NH_2, \; R_2 = H$

A chemist on the project proposed synthesizing compound XIII. He recalls that at the meeting at which he suggested it, I was the only one that thought it was a good idea. Presumably I was thinking about diversity in physical properties which this would certainly add.

In contrast to the lead, XIII is a high-ceiling diuretic and very potent. Subsequent molecular modification led to XIV, which is even more potent. This series was explored for several years with many interesting compounds produced (Plattner et al. 1982; Martin and Kim 1984).

3.4.3 Ristocetin and Vancomycin (1987)

The antibacterial compound ristocetin, XV, exerts its effect by binding to the growing bacterial cell wall. In the early 1980s Abbott started a project based on the 3D structure of the complex between ristocetin and acetyl-lys-D-ala-D-ala, a model of a peptide of the cell wall. The binding constant for this reaction is ca 10^{-5} M. We reasoned that the structure of the complex should help us design lower molecular weight and more water-soluble ristocetin analogs that might be orally active or have more activity against gram-negative organisms. Extensive NMR studies and molecular dynamics calculations were done in support of the chemical synthesis effort.

XV

I did no calculations for the project, but attended the biweekly meetings to follow the progress. Because of the low affinity of the complex, I suspected that ristocetin was perhaps not such a good lead as we hoped. Hence, when I heard about GRID (Goodford 1985), I was very eager to see if it would help me design de novo a good ligand for lys-D-ala-D-ala. It produced very sensible potential interaction sites. However, I abandoned this de novo design rather quickly since the chemistry of the project was committed to mimicking at least part of the structure of ristocetin and I also realized that it would not be easy to design a compound that binds to the peptide and is easy to make. The experience with GRID also put me in a good position to appreciate CoMFA (Cramer et al. 1988; Patterson et al. 1988) when I heard about it a few years later.

The biochemists measured the ability of acetylated di- and tripeptides to compete for the bacterial cell wall in binding to ristocetin. They also measured the affinity of these compounds for vancomycin. I suspected that the low binding affinity of ristocetin for its target was a result of a poor fit between some part of the antibiotic and the peptide. Hence, I suggested that Ki Kim derive QSAR equations for these two datasets. He found good equations for both (Kim et al. 1989). From the equations I estimated that if everything in the complex were ideal, the affinity of ristocetin should be 6 kcal/mole greater than that observed. A simple molecular graphics evaluation of the structure of the complex that best fit the NMR data revealed a steric bump at the position of D-ala$_2$. The only way to relieve this would be to drastically change the structure of ristocetin. I concurred with the decision made a few months later to abandon the project.

3.5 Examples of Collaborations with Statisticians that Led to Improvements in Computer Drug Design Methodology

3.5.1 Discovery of Discriminant Analysis (1973)

I was dismayed by the question of a synthetic chemist who wanted to know why QSAR cannot be applied to data that we get from screening (in those days: ++, +, and 0)? Fortunately, I talked frequently to a biostatistician who tried to broaden my statistical perspective. He invited me to an intensive course in multivariate statistics that he had set up for his colleagues. When I heard about discriminant analysis, I knew it would be interesting to apply to QSAR. A few months later I started to apply it with good QSAR success.

This example shows two important points. First, it is worthwhile to listen to critics of computer methods. Only by keeping in mind the deficiencies of our strategies will we recognize a chance to extend the method further. Second, it is important to talk repeatedly to people of diverse disciplines about the scientific issues that you face. They might be able to head you in a fruitful direction.

3.5.2 Quick Calculation of Molecular Surfaces (1984)

After the success with discriminant analysis I had no trouble convincing the management of the biostatisticians that I had interesting problems to work on. Now they assign their best preclinical statistician to work with me roughly half-time.

As we started molecular modeling, the input of traditional biostatistics into our work was less important than it had been when our emphasis was QSAR. I was fortunate, therefore, that my assigned collaborator had a strong training in geometry.

For our molecular graphics program we wanted an algorithm that calculated molecular surfaces more quickly than those currently in use. We also wanted to display the intersection or difference in the volumes occupied between two groups of compounds – active versus inactive, for example. Well before it was published by others, this statistician programmed a simple lattice-based algorithm that accomplishes both requirements. Since the difference in volumes is interesting only if it is significant, she programmed it so that the thickness of the volume differences must be larger than some default, usually 0.5 Å. She also added an option to cluster the points in the volume differences so that we could color them differently, display only certain clusters, or calculate the volumes of individual clusters. Not only did my collaborator solve the immediate problem, but she taught me the value of looking at molecular structures from the perspective of a lattice – a valuable prelude to CoMFA.

3.5.3 Minima and Maxima in ESP Surfaces as a Prelude to DISCO (1984)

We also wanted to be able to superimpose compounds according to the maxima and minima in their electrostatic potential surface. This statistician developed an algorithm that identifies the locations (and energies) of these extreme values, by interpolation if necessary. Imagine our amusement when we discovered that these locations are often well approximated by extending lone pairs or hydrogen-bonding hydrogens to the surface! While we used the program quite extensively for a while, the insights gained from it were critical in our assignment of site points for our pharmacophore mapping program DISCO developed almost a decade later (Martin et al. 1993a).

3.6 Suggestions of Management Strategies to Increase the Collaboration of Synthetic and Computational Chemists

3.6.1 Encourage Collaboration Without Forcing It

Clearly, management would like to have scientists working to solve the most important problems of the organization. Hence, they are inclined to assign computational chemists to projects with the most pressing needs. Making such assignments can be a successful strategy. However, it can happen that the synthetic chemists interpret the assignment as a lack of management confidence that they can solve the problems on their own. They might also be skeptical of the computer methods and/or their appropriateness to solving the current problem. In such an atmosphere, it is very difficult for a true collaboration to develop.

Management can also encourage collaboration by monitoring that the parties are paying enough attention to it. Successful collaborations are characterized by intense and frequent interactions. Even if both people are spending time on activities of mutual interest, if they do not talk about them with each other, then the benefits of a collaboration are not realized.

3.6.2 Educate the Potential Collaborators
to Increase Respect for Each Other's Disciplines

It might seem like a good idea to work together on some problem, but unless both parties are willing to learn something about the other's approaches, frustration is likely to ensue. All approaches take time and expertise to apply, have limitations to their scope, and even have different definitions of success. Until the co-workers truly understand each other's viewpoints, it will be difficult for them to develop a true collaboration.

Sometimes the request for computational help on a problem is a more of a challenge than a suggestion for collaboration. This can happen if a synthetic chemist, out of frustration, hopes that modeling will solve a long-standing problem, such as replacement of a functionality that is considered to be part of the essential pharmacophore. The computer person then faces a difficult design problem without a knowledge of the existing SAR. Realistically, no sensible suggestions can be made until the SAR is understood and perhaps a CoMFA developed. However, the synthetic chemist may be desperate because to his mind every sensible compound has already been made and may also be unwilling to wait until a reasonable CAMD effort matures. Additionally, if novel compounds are suggested, it is easy to dismiss them because they "look odd" and the chemist is already convinced that the goal is impossible to meet. Such self-fulfilling prophecies can be very frustrating to the CAMD person and lead to disappointment for the synthetic chemist.

A complementary problem can arise if the CAMD person expects that every suggested compound will be made. When suggestions are rejected because the compounds are "impossible to make" the CAMD person may feel that the synthetic chemist did not try hard enough to come up with a synthesis. At the same time, the synthetic chemist may feel that the CAMD person expects too much experimental information.

3.7 Disadvantages of Collaborations

It takes time for two or more people to reach a common viewpoint. It might look to outsiders that nothing is happening while the discussions are taking place. Additionally, if one of the collaborators is domineer-

ing, the collaboration might lead to work at the lowest common denomi-
nator rather than synergistic.

Mutual control can be hard to manage from the outside. If a collaboration reaches a successful completion, who is given the "honors"? How
does a manager evaluate the contribution of the individuals? Sometimes
it looks as if each member of the collaboration is a technician for the
other. This problem is compounded when the collaborating individuals
report to different managers, as often is the case.

3.8 Summary

In this report I have illustrated the power of collaboration: the results far
outstrip the initial expectations and often lead both the medicinal and
computational chemistry research in unexpected directions. A collaboration requires a close working relationship between the parties and
hence presents a challenge to management.

References

Borman S (1989) Software adds new dimension to structure searching. Chem
    Eng News 67:28–32
Cramer RD III, Patterson DE, Bunce JD (1988) Comparative Molecular Field
    Analysis (CoMFA). 1. Effect of shape on binding of steroids to carrier pro-
    teins. J Am Chem Soc 110:5959–5967
DeNinno MR, Schoenleber R, MacKenzie R, Britton DR, Asin KE, Briggs C,
    Trugman JM, Ackerman M, Artman L, Bedzarz L, Bhatt R, Curzon P, Gomez
    E, Kang CH, Stittsworth J, Kebabian JW (1991) A-68930: a potent agonist se-
    lective for the dopamine D1 receptor. Eur J Pharmacol 199:209–219
Goodford P (1985) A computational procedure for determining energetically
    favorable binding sites on biologically important macromolecules. J Med
    Chem 28:849–857
Kim K-H, Martin Y, Otis E, Mao J (1989) Inhibition of the 125I-labelled ris-
    tocetin binding to micrococcul luteus cells by the peptides related to bac-
    terial cell wall mucopeptide precursors: quantitative structure-activity rela-
    tionships. J Med Chem 32:84–93
Martin YC (1990) Computer-aided design of potentially bioactive molecules
    by geometric searching with ALADDIN. Tetrahedron Comput Methodol
    3:15–25

Martin YC (1992) 3D database searching in drug design. J Med Chem 35:2145–2154

Martin YC (1994) Molecular modeling in the design of potent and selective D1 dopaminergic agonists for potential use in parkinsonism. ACS Satellite TV Seminar: molecular modeling: the small-molecule approach. American Chemical Society, Washington DC

Martin YC, DeLazzer J (in press) Family: a computer program that compares 3D structures based on interpoint distances: application to sorting hits from 3D database searching. J Comput Aided Mol Design

Martin YC, Kim KH (1984) Application of theoretical drug design methodology to a series of diuretics. Drug Inform J 18:95–113

Martin YC, Van Drie JH (1993) Identifying unique core molecules from the output of a 3D database search. Chemical structures 2. The international language of chemistry. Springer, Berlin Heidelberg New York, pp 315–326

Martin YC, Danaher EB, Weininger AM, Weininger D (1986) AIMM: 3-D models from 2-D MACCS connection tables–use of the GENIE target language to specify rules for structure building. Abstracts for Molecular Graphics Society meeting

Martin YC, Danaher EB, May CS, Weininger D (1988) MENTHOR, a database system for three-dimensional structures and associated data searchable by substructure alone or combined with geometric properties. J Comput Aided Mol Design 2:15–29

Martin YC, Kebabian JW, MacKenzie R, Schoenleber R (1991) Molecular modeling-based design of novel, selective, potent D1 dopamine agonists. QSAR: rational approaches on the design of bioactive compounds. Elsevier, Amsterdam, pp 469–482

Martin YC, Kim K-H, Bures MG (1992) Computer-assisted drug design in the 21st century. Medicinal chemistry in the 21st century. Blackwell, Oxford, pp 295–317

Martin YC, Bures MG, Danaher EA, DeLazzer J, Lico I, Pavlik PA (1993a) A fast new approach to pharmacophore mapping and its application to dopaminergic and benzodiazepine agonists. J Comput Aided Mol Design 7:83–102

Martin YC, Lin CT, Wu J (1993b) Application of CoMFA to the design and structural optimization of D1 dopaminergic agonists. 3D QSAR in drug design. Theory methods and applications. Escom, Leiden, pp 643–660

MedChem Software Manual Release 3.54 (1989) Claremont, California, Daylight Chemical Information Systems

Pearlman RS, Rusinko A III, Skell JM, Balducci R, McGarity CM (1988) Concord. St Louis, Missouri 63944, Tripos Associates, 1699 S. Hanley Road, Suite 303

Plattner JJ, Lee CM, Horrom BW, Ours CW, Martin YC, Pernet AG (1982) Benzylamine diuretics. A unique class of (aryloxy)acetic acid derivatives. Abstract papers of the American Chemical Society, Sept 1982

Schoenleber RW, Kebabian JW, Martin YC, DeNinno MP, Perner RJ, Stout DM, Hsiao C-NW, DiDomenico S Jr, DeBernardis JF, Basha FZ, Meyer MD, De B (1990) Dopamine agonists. US patent 4,963,568

Van Drie JH, Weininger D, Martin YC (1989) ALADDIN: an integrated tool for computer-assisted molecular design and pharmacophore recognition from geometric, steric, and substructure searching of three-dimensional molecular structures. J Comput Aided Mol Design 3:225–251

4 Structure-Based Ligand Design

K. Gubernator, C. Broger, D. Bur, D. M. Doran, P. R. Gerber,
K. Müller, and T. M. Schaumann

4.1 Introduction

Structure-based ligand design has matured over the last decade. This process has been driven by an exploding wealth of biostructural information and by a detailed understanding of the mechanism of action of a wide range of biochemical systems. While 10 years ago only few structures relevant for human pathology were known, today most newly initiated pharmaceutical research projects have at least some relation to known three-dimensional (3D) structures. Examples of research areas where solid biostructural information is available are: Serine proteases (e.g., elastase (Powers et al. 1990)), coagulation factors (Banner 1993), metalloproteinases (e.g., collagenase (Borkakoti et al. 1994), astacin (Bode et al. 1992)), aspartyl proteases [e.g., renin (Dealwis et al. 1994), HIV protease (Lam et al. 1994)], serine esterases [several lipases (Winkler et al. 1990; Cygler et al. 1993)], acetylcholine esterase (Sussman et al. 1991; Gubernator et al. 1993), dihydrofolate reductase (Oefner et al. 1988), tymidylate synthetase (Shoichet et al. 1993), major histocompatibility complex (Silver et al. 1992), cyclosporin-cyclophilin (Ke et al. 1994; Spitzfaden et al. 1994), hemagglutinin (Von Itzstein et al. 1993), and protein–DNA interaction [transcription factors (Harrison 1991), nucleases (Winkler 1993)].

The traditional ligand design approach typically consists of (1) concentration of related structure and sequence information into a consistent 3D view of the biochemical knowledge about the target system, (2) forcefield-assisted manual construction of models of experimentally inaccessible complexes and reaction pathway intermediates, and (3) modeling of novel inhibitor structures inside the protein cavity guided by the synthetic knowledge of a chemist.

Recent additions to the methodological repertoire of the design process are 3D database searches and denovo design tools that complement the traditional approach.

In the following two recent enzyme-based projects will be discussed that include biostructure-derived design components: Thrombin and β-lactamase inhibition. In addition, an example of a denovo design tool will be presented.

4.2 Thrombin Inhibitors

Thrombin, as well as other coagulation factors, is a trypsinlike serine protease. The mechanism of action of this class of enzymes is well understood based on a large number of structural studies (Huber et al. 1974). These include crystallographic work with substrates and derivatives, with mechanism-based inhibitors and with noncovalent inhibitors, as well as nuclear magnetic resonance (NMR) studies with substrate peptides (Banner and Hadvary 1991; Stubbs et al. 1992; Feng et al. 1989). To summarize, the scheme of the substrate cleavage by trypsin and a model of the transition state of the acylation reaction are shown in Fig. 1. The transition state is characterized by a tetrahedral carbon covalently linked to the catalytic serine Og, a negatively charged oxyanion stabilized by two hydrogen bonds to two backbone NH groups, and a protonated histidine which has received the proton from the serine-OH. In the next step, the leaving group will be activated by a proton transfer from the histidine to the amide nitrogen atom; the former amide bond will break and a serine ester will be formed. The deacyla-

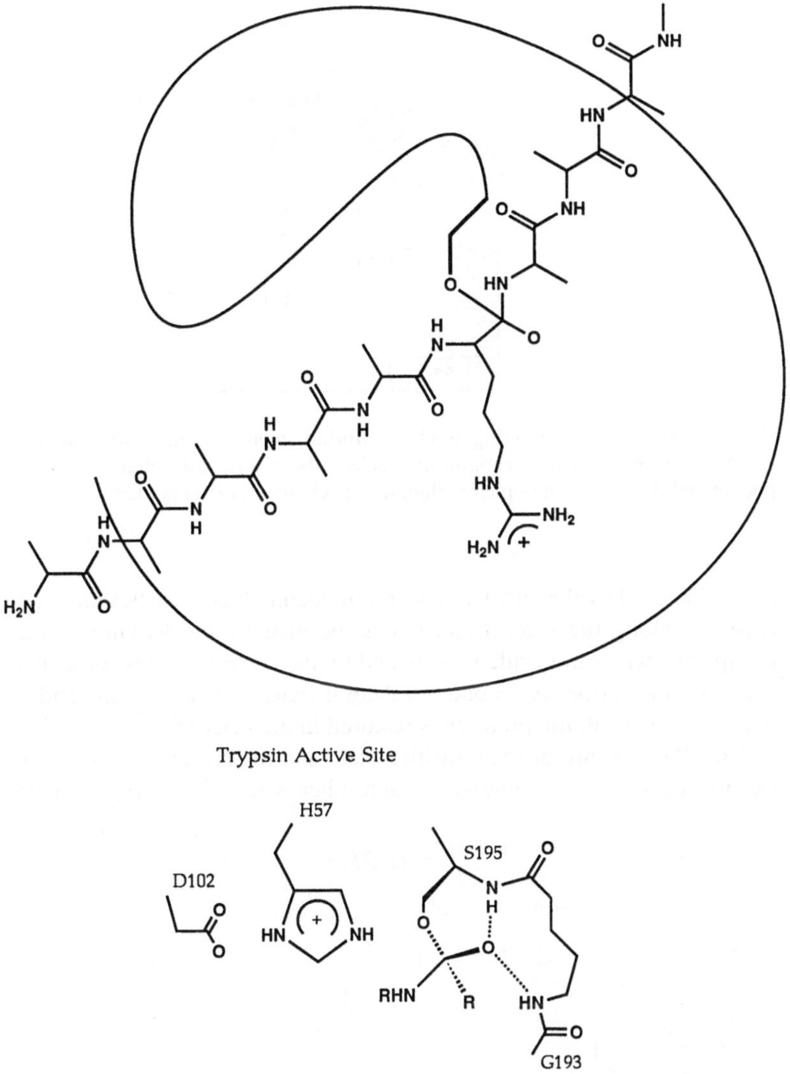

Fig. 1. Schematic representation of the cleavage of a peptide by trypsin at the amide bond after an arginine residue

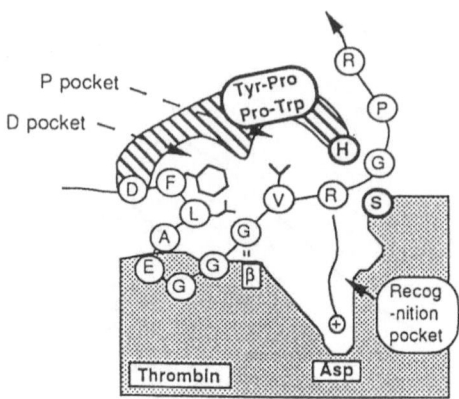

Fig. 2. Cartoon of the binding mode of fibrinogen in thrombin with the arginine side chain in the recognition pocket, the valine side chain in the P pocket and the leucine and phenylalanine side chains in the D pocket

tion is initiated by the attack of a water molecule. It approaches the ester carbonyl along the same trajectory as defined by the leaving amine group. The water molecule is activated by the same basic histidine that activated the serine. A second tetrahedral transition state forms and it decays to the acid; the protease is restored in the active form.

The 3D structure of the transition state and the protein environment exhibit ideal complementarity to each other, which is the prerequisite

MD - 805 NAPAP

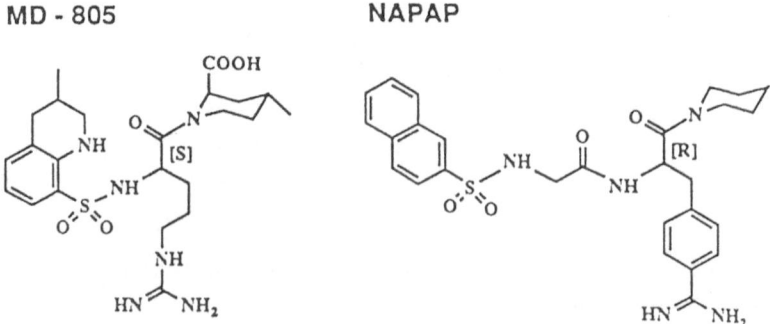

Fig. 3a,b. Formulae of the MD805 (**a**) and NAPAP (**b**) thrombin inhibitors

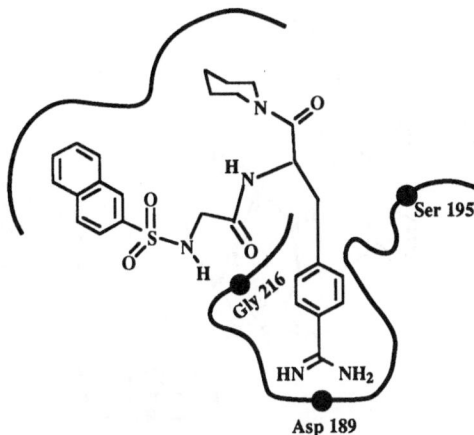

Fig. 4. Schematic representation of the experimentally determined binding mode of NAPAP to thrombin

for the observed rate enhancement of the hydrolytic reaction. The specificity for cleavage after arginine or lysine is mediated by a recognition pocket with an aspartate side chain at the bottom. This pocket accommodates the positively charged side chain and forms a salt bridge with the carboxylate of the aspartic acid.

In thrombin, the specificity extends far beyond the residue preceding the cleavage site. NMR (Feng et al. 1989) and X-ray studies (Banner and Hadvary 1991; Martin et al. 1992) revealed that three hydrophobic residues in fibrinogen two, eight, and nine residues before the cleavage site, respectively, occupy unique hydrophobic pockets (Fig. 2). These hydrophobic pockets are formed by a unique additional loop Tyr-Pro-Pro-Trp above the active site.

The binding mode of two previously known inhibitors MD805 and NAPAP (Fig. 3) as revealed in the X-ray structure of their complexes with thrombin (Banner and Hadvary 1991; Martin et al. 1992) has some unexpected features: The basic moiety binding in the specificity pocket is directly connected to hydrophobic portions occupying the two hydrophobic pockets; there is no interaction with the catalytic serine. The inhibitors thus do not bind substrate-like, but bypass the catalytic site and preferentially interact with the hydrophobic pockets (Fig. 4). In the

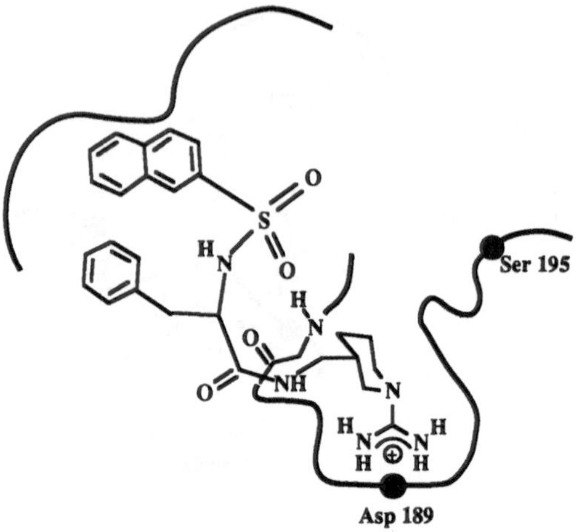

Fig. 5. Schematic representation of the experimentally determined binding mode of the D-Phe derivative to thrombin

complex with fibrinogen, the same pockets are occupied by the hydrophobic sidechains preceding the cleavage site.

Based on these findings, two new classes of potent thrombin inhibitors have been discovered by a three-step process (Banner et al. 1993, 1994). First, a collection of small, basic molecules have been tested in

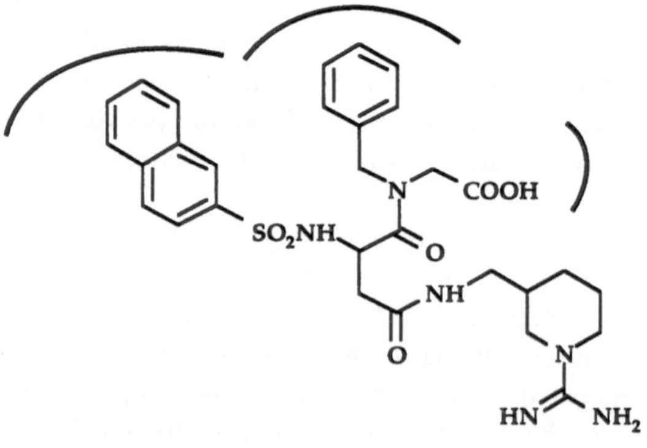

Fig. 6. Schematic representation of the experimentally determined binding mode of the aspartate-benzyl-glycine derivative to thrombin

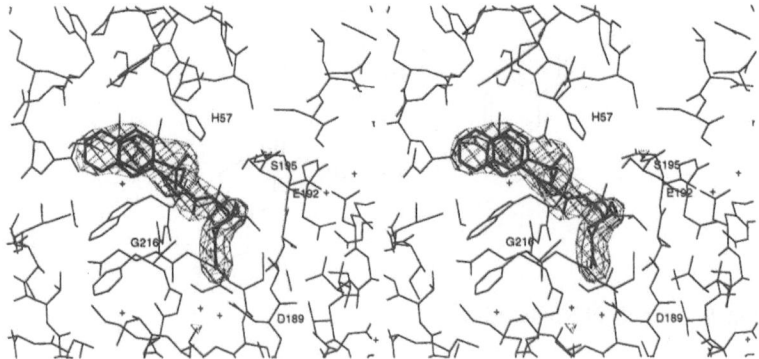

Fig. 7. Omit map difference electron density for the D-Phe derivative

thrombin and trypsin assays; N-amidinopiperidine has been selected as being more active than benzamidine and slightly selective for thrombin. Secondly, substituents attached to the 3-position of amidinopiperidine containing two hydrophobic moieties mediate low nanomolar activity (Schmid et al. 1990; Fig. 5). Replacement of the d-aminoacid middle portion by a side chain-connected l-aspartate finally leads to picomolar thrombin inhibitors which are highly selective (Fig. 6; Hilpert et al. 1994).

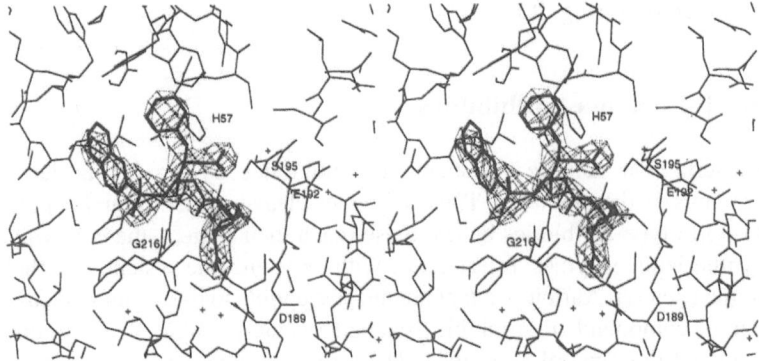

Fig. 8. Omit map difference electron density for the aspartate-benzyl-glycine derivative

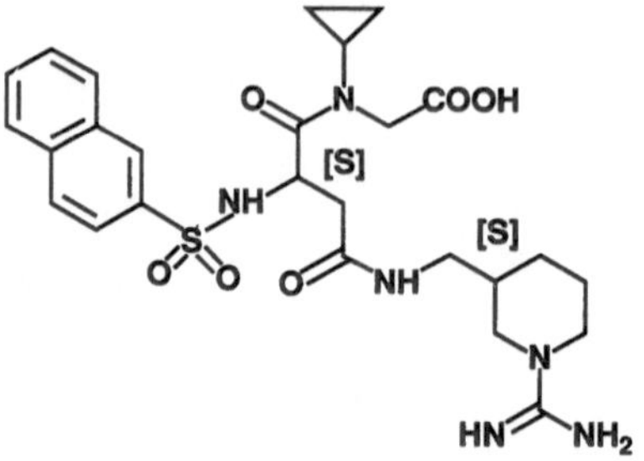

Fig. 9. Formula of Ro-46–6240

Several of these compounds have been studied by X-ray crystallography of complexes with human thrombin; two representative examples are shown in Figs. 5,7 and 6,8, respectively.

Interestingly, they exhibit two different binding modes where the occupancy of the two neighboring hydrophobic pockets is distinctly different. In the case of the d-aminoacid derivatives, the naphthyl portion occupies the rear side of the pockets toward histidine 57; in the aspartate case, the substituent on the glycine nitrogen occupies the inner pocket and naphthylsulfonyl unit the outer one.

From the latter class of compounds, the N-cyclopropylglycin derivative Ro-46–6240 has been selected for clinical development as a possible intravenous antithrombotic drug based on pharmacokinetic considerations (Fig. 9.)

4.3 β-Lactamase Inhibitors

Bacteria can acquire resistance to antibiotics by producing large amounts of β-lactamases. These enzymes cleave penicillin and cephalosporin type antibiotics before these reach their targets, the cell wall-synthesizing enzymes. Interference with these enzyme systems is lethal to the bacteria. Antibiotics that are not susceptible to β-lactamase cleavage or compounds that inhibit β-lactamases are possible means to circumvent this type of resistance. The target enzymes of β-lactam antibiotics all belong to the same superfamily and most likely share the same catalytic mechanism. Therefore, classes of inhibitors might be

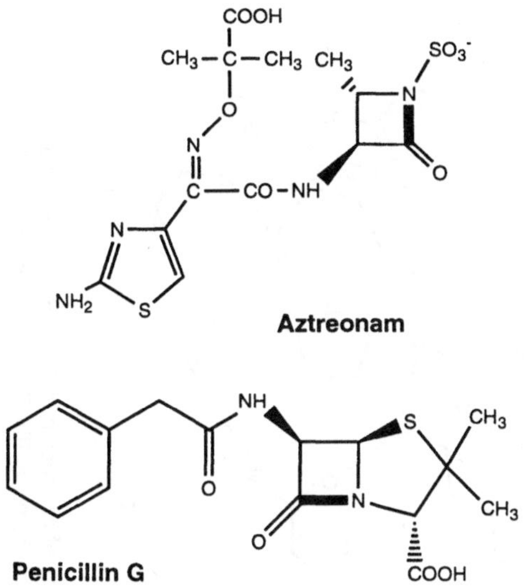

Fig. 10. Formulae of aztreonam and penicillin G

discovered that inhibit several or all members of that family of enzymes. This could open new opportunities for advances in antibacterial therapy.

As a first step in this direction, the structure of the class C β-lactamase from *Citrobacter freundii* and its complex with the monobactam antibiotic aztreonam has been studied (Oefner et al. 1990). In the complex structure the reaction between enzyme and inhibitor has been trapped at the stage of the acyl enzyme, after the ring-opening acylation. The arrangement in space of the functional groups around the serine is similar to that in trypsin or thrombin. However, the serine activation is thought to be brought about by a tyrosine flanked by two lysine NH^{3+} which as a group are deficient by one proton (Figs. 10,11). Based on this experimental structure, a hypothetical model of the intermediate acyl enzyme stage of penicillin G was built. The orientation of the acylamino side chain and of the negatively charged group ($-SO^{3-}$ in aztreonam, $-COO^-$ in penicillin) are similar, but the substituents that could be in the way of the incoming water are different. This structural detail could help to rationalize the observed differences in deacylation rate, which is slow for aztreonam and fast for penicillin G: In penicillin G, the confor-

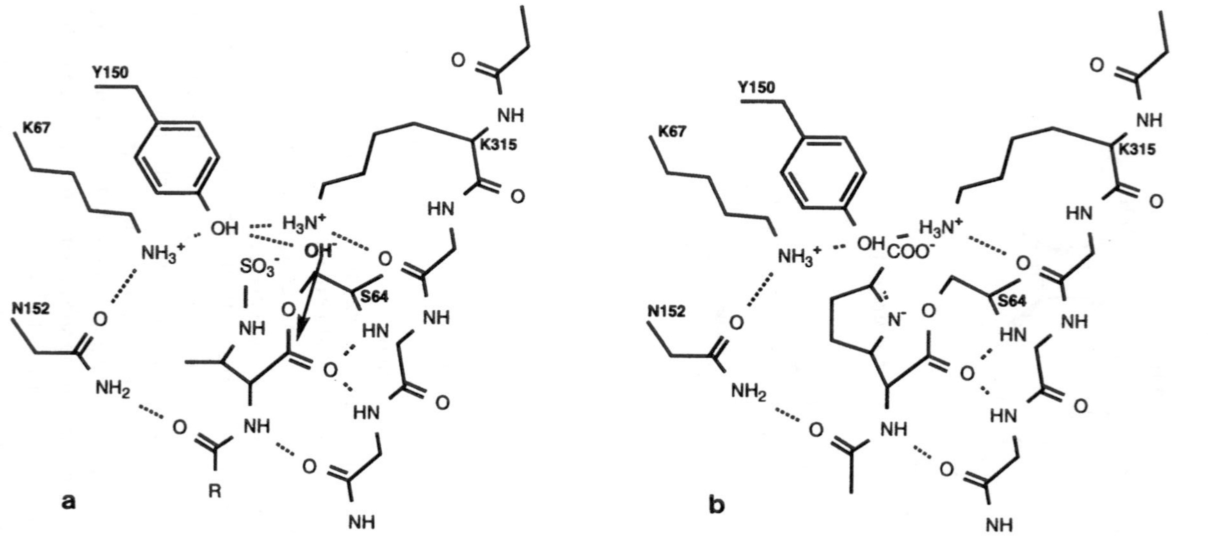

Fig. 11a,b. Sketch of the experimentally observed acyl enzyme complex of aztreonam in *C. freundii* β-lactamase (**a**) with an incoming water which collides with the sulfonyl group; in **b** the model of the penicillin G acyl enzyme intermediate is shown. The outward rotation of the thiazolidine ring gives way to a deacylating water

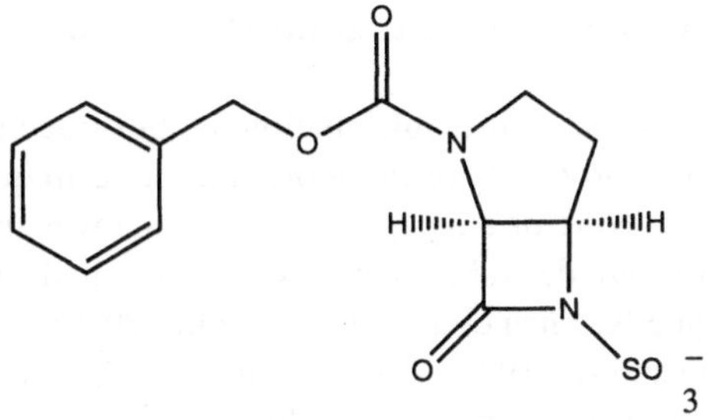

Fig. 12. Formula of Ro-44-4454

mation of the 3-4 bond resulting from the β-lactam ring opening is doubly eclipsed and relief of this strain can only be achieved through outward rotation of the thiazolidine ring. In the case of aztreonam, by contrast, an unstrained conformation is accessible through slight inward rotation (as observed in the X-ray structure) because of the opposite stereochemistry at position 4.

Based on these structural concepts of the deacylation mechanism, a novel class of inhibitors was suggested in which an additional ring bridging the 3-4 C–C bond restricts the rotation around this bond and, as a consequence, prevents deacylation. The first example in a series of compounds in this class was found to be a very potent inhibitor of class C β-lactamases (Charnas et al. 1991; Fig. 12). It was shown that inhibition was indeed caused by an acyl enzyme intermediate stable to hydrolysis.

Interestingly, in the case of the class A β-lactamases, exemplified by the known structure of the *Bacillus licheniformis* enzyme (Moews et al. 1990), an alternative deacylation pathway may be available. A water molecule could approach the serine ester from the opposite side; it would be activated by a glutamic acid side chain which is unique to the class A enzymes.

These mechanistic insights may have an impact on ongoing research directed towards the discovery of β-lactamase inhibitors to be combined with potent, third-generation cephalosporins or, as a long-term goal, towards β-lactamase-insensitive antibiotics.

4.4 LEGO, A New Denovo Ligand Design Tool

The sequence and structure information in biostructure research is growing exponentially. It happens more and more frequently that the structure of an enzyme or a receptor protein is known and suggestions for new inhibitors are desirable. This is where so called "de novo design tools" are applied (Caflish et al. 1993; Boehm 1992; Lauri and Bartlett 1994; Nishibata and Itai 1993; Von Itzstein et al. 1993; Moon and Howe 1991). These either build up candidate ligands from atoms or fragments or they search databases of existing structures for complementary molecules.

Our LEGO tool for systematic exploration of protein cavities combines both approaches. The procedure is outlined in Fig. 13. In the first step, small molecular fragments such as benzene, imidazole, or methylguanidine are positioned on a multitude of surface points in a number of orientations around the active site or binding pocket (for thrombin we choose about 120 surface points spaced by at least 1A and selected 24 initial orientations of the fragment molecule at each point). Each fragment is energy minimized to its next local minimum against a rigid protein environment; similar and symmetry-related structures as well as high energy minima are discarded. These favorably positioned fragments can be connected by appropriate linker fragments. These consists of a one to four atom chain and usually contain at least one synthetically well accessible bond, e.g., amide, sulfonamide, ester, or carbamate. All possible conformations of the linkers are stored in a linker library together with the geometrical parameters describing the linker exit vectors to be combined with the positioned fragments.

The resulting ligands are compatible with the surrounding protein, synthetically accessible and conformationally unstrained. The linking step can be repeated and more complex ligands can thus be constructed.

It is planned to extend the scope of the methods by adding a scoring function (e.g. Boehm 1992) that allows estimation of the relative binding constants and thus the selection of the most promising candidates for synthesis.

A typical LEGO run in the thrombin active site cavity with twelve fragments requires about 1 h on an eight processor SGI ONYX evaluating 2400 orientations for each fragment using the MAB force field (Gerber and Müller 1995). A few hundred positions are kept per frag-

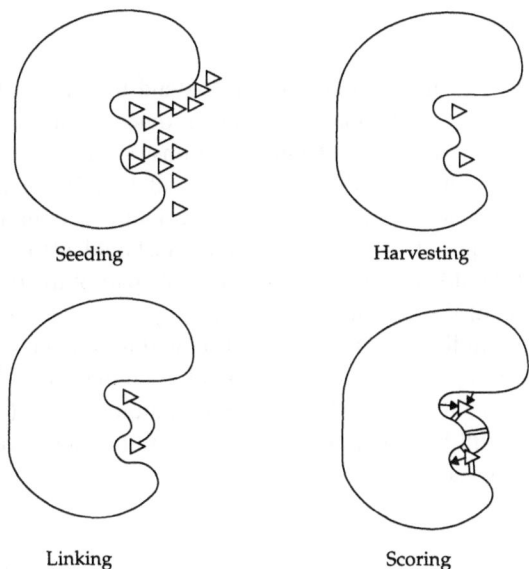

Fig. 13. Illustration of the stepwise construction of novel ligands in the LEGO method

ment. Linking of the docked fragments with a library of 50 linkers in a total of 200 conformations requires a few seconds per fragment set; a complete run linking every fragment position to all other fragment sets and exploring about 50 million possibilities requires 1–2 min (R4400 ONYX, one processor). The LEGO method has been implemented within the MOLOC modeling system (Müller et al. 1988, 1989; Gerber 1992; Gerber et al. 1988; Gerber and Müller 1994).

Medicinal chemistry assisted by methods such as LEGO offers promise to considerably speed up the lead discovery process. They can be customized to a specific target or an enzymatic class in terms of size and property distribution of molecules. They also complement emerging parallel and combinatorial synthesis approaches well and could, combined with automated screening, constitute a drug discovery multi-cycle.

4.5 Conclusion

In favorable cases it is possible to understand the drug action at the target enzyme in considerable detail; this provides a firm basis for the rationalization of structure–affinity relationships and the design of novel potent inhibitors. To this end a close collaboration of scientists in chemistry, biochemistry, pharmacology, structure determination, and modeling is required; such a collaboration should start as early as possible and should persist over a considerable part of the project's life cycle. In view of the current general tendency to shorten project duration, such a multidisciplinary collaboration will have to be well coordinated and will require considerable flexibility on the part of all participants. New communication tools and information systems (Schatz and Hardin 1994) will be required to assist in the tight interaction between all participating scientists.

References

Ackermann J, Banner DW, Gubernator K, Hilpert K, Schmid G, (1992) European patent 559046

Banner DW, Hadvary P (1991) Crystallographic analysis at 3.0-A resolution of the binding to human thrombin of four active site-directed inhibitors. J Biol Chem 266:20085–20093

Banner DW, Ackermann J, Gast A, Gubernator K, Hadvary P, Hilpert K, Labler L, Müller K, Schmid G, Tschopp T, van de Waterbeemd H, Wirz B (1993) Serine proteases: 3D structures, mechanism of action and inhibitors. In: Testa B, Kyburz E, Fuhrer W, Giger R (eds) Perspectives in medicinal chemistry. VHCA, Basel

Bode W, Gomis-Rueth FX, Huber R, Zwilling R, Stoecker W (1992) Structure of astacin and implications for activation of astacins and zinc-ligation of collagenases. Nature 358:164–167

Boehm HJ (1992) The computer program LUDI: a new method for the de novo design of enzyme inhibitors. J Comput Aided Mol Des 6:61–78

Borkakoti N, Winkler FK, Willams DH, D'Arcy A, Broadhurst MJ, Brown PA, Johnson WH, Murray EJ (1994) Structure of the catalytic domain of human fibroblast collagenase complexed with an inhibitor. Nature Struct Biol 1:106–110

Caflish A, Miranker A, Karplus M (1993) Multiple copy simultaneous search and construction of ligands in binding sites: application to inhibitors of HIV-1 aspartatic proteinase. J Med Chem 36:2142–2167

Charnas RL, Gubernator K, Heinze I, Hubschwerlen C (1991) European patent 508234

Cygler M, Schrag JD, Sussman JL, Harel M, Silman I, Gentry MK, Doctor BP (1993) Relationship between sequence conservation and three-dimensional structure in a large family of esterases, lipases, and related proteins. Protein Sci 2:366–382

Dealwis CG, Frazao C, Badasso M, Cooper JB, Tickle IJ, Driessen H, Blundell TL, Murakami K, Miyazaki H, Sueiras-Diaz J, Jones DM, Szelke M (1994) X-ray analysis at 2.0 Å resolution of mouse submaxillary renin complexed with a decapeptide inhibitor CH-66, based on the 4–16 fragment of rat angiotensinogen. J Mol Biol 236:342–360

Feng N, Konishi Y, Frazier RB, Scheraga HA (1989) High-resolution NMR studies of fibrinogen like peptides in solution: interaction of thrombin with residues 1–23 of the A.alpha. chain of human figrinogen. Biochemistry 28:3082–3094

Gerber PR (1992) Peptide mechanics: a force field for peptides and proteins working with entire residues as smallest units. Biopolymers 32:1003–1017

Gerber PR, Müller K (1995) MAB: a generally applicable molecular force field for structure modelling in medicinal chemistry. J Comput Aided Mol Design (in press)

Gerber PR, Gubernator K, Müller K (1988) Generic shapes for the conformational analysis of macrocyclic structures. Helv Chim Acta 71:1429–1441

Gubernator K, Ammann HJ, Broger C, Doran DM, Gerber PR, Müller K, Schaumann TM (1993) The mechanism of action and inhibition of pancreatic lipase and acetylcholinesterase: a comparative modeling study. Mol Simul 10:211–223

Harrison SC (1991) A structural taxonomy of DNA-binding domains. Nature 353:715–719

Hilpert K, Ackermann J, Banner DW, Gast A, Gubernator K, Hadvary P, Labler L, Müller K, Schmid G, Tschopp T, van de Waterbeemd H, Wirz B (1994) Design and synthesis of potent and highly selective thrombin inhibitors. J Med Chem 37:3889–3901

Huber R, Kukla D, Bode W, Schwager P, Bartels K, Deisenhofer J, Steigemann W (1974) Structure of the complex formed by bovine trypsin and bovine pancreatic trypsin inhibitor. J Mol Biol 89:73–101

Ke H, Mayrose D, Belshaw PJ, Alberg DG, Schreiber SL, Chang ZY, Etzkorn FA, Ho S, Walsh CT (1994) Crystal structures of cyclophilin A complexed with cyclosporin A and N-methyl- 4-[(E)-2-butenyl]-4,4-dimethylthreonine cyclosporin A. Structure 2:33–44

Lam PYS, Jadhav PH, Eyermann CJ, Hodge CN, Ru Y, Bacheler LT, Meek JL, Otto MJ, Rayner MM, Wong YN, Chang C-H, Weber PC, Jackson DA, Sharpe TR, Erickson-Viitanen S (1994) Rational design of potent, bioavailable, nonpeptide cyclic ureas as HIV protease inhibitors. Science 263:380–384

Lauri G, Bartlett PA (1994) CAVEAT: a program to facilitate the design of organic molecules. J Comput Aided Mol Design 8:51–66

Martin PD, Robertson W, Turk D, Huber R, Bode W, Edwards BFP (1992) The structure of residues 7–16 of the Aa-chain of human fibrinogen bound to bovine thrombin at 2.3.-angstroem resolution. J Biol Chem 267:7911–7920

Moews PC, Knox JR, Dideberg O, Charlier P, Frere JM (1990) Beta lactamase of bacillus licheniformis 749/C at 2 angstrom resolution. Proteins 7:156–171

Moon JB, Howe WJ (1991) Computer design of bioactive molecules: a method for receptor-based de novo ligand design. Proteins 11:314–328

Müller K, Ammann HJ, Doran DM, Gerber PR, Gubernator K, Schrepfer G (1988) Complex heterocyclic structures – a challenge for computer-assisted molecular modeling. Bull Soc Chim Belg 97:655

Müller K, Ammann HJ, Doran DM, Gerber PR, Gubernator K, Schrepfer G (1989a) Use of computer modeling and structural databases in pharmaceutical research. In: van der Groot H, Domany G, Pallos L, Timmerman H (eds) Trends in medicinal chemistry. Elsevier, Amsterdam

Müller K, Ammann HJ, Doran DM, Gerber PR, Gubernator K, Schrepfer G (1989b) Computer modeling and structural databases in pharmaceutical research. In: Richards RWG (ed) Computer-aided molecular design. IBC, London

Nishibata Y, Itai A (1993) Confirmation of usefulness of a structure construction program based on three-dimensional receptor structure for rational lead generation. J Med Chem 36:2921–2928

Oefner C, D'Arcy A, Winkler FK (1988) Crystal structure of human dihydrofolate reductase complexed with folate. Eur J Biochem 174:377–385

Oefner C, D'Arcy A, Daly JJ, Gubernator K, Charnas RL, Heinze I, Hubschwerlen C, Winkler FK (1990) Refined crystal structure of Beta-lactamase from Citrobacter freundii indicates a mechanism for Beta-lactam hydrolysis. Nature 343:284–288

Powers JC, Oleksyszyn J, Narasimhan SL, Kam C-M, Radhakrishnan R, Meyer EF Jr (1990) Reaction of porcine pancreatic elastase with 7-substituted 3-alkoxy-4-chloro-isocoumarins: design of potent inhibitors using the crystal structure of the complex formed with 4-chloro-3-ethoxy-7-guanidinoisocoumarin. Biochemistry 29:3108–3118

Schatz BR, Hardin JB (1994) NCSA mosaic and the world wide web: global hypermedia protocols for the internet. Science 265:895–901

Schmid G, Ackermann J, Banner DW, Gubernator K, Hadvary P, Hilpert K, Labler L, Müller K, Tschopp T, Wessel HP, Wirz B (1990) European patent 0468231

Shoichet BK, Stroud RM, Santi DV, Kuntz ID, Perry KM (1993) Structure-based discovery of inhibitors of thymidylate synthase. Science 259:445–1450

Silver ML, Guo H, Strominger JL,Wiley DC (1992) Atomic structure of a human MHC molecule prsenting an influenza virus peptide. Nature 360:367–369

Spitzfaden C, Braun W, Wider G, Widmer H, Wuthrich K (1994) Determination of the NMR solution structure of the cyclophilin A-cyclosporin A complex. J Biomol NMR 4:463–482

Sussman JL, Harel M, Frolow F, Oefner C, Goldman A, Toker L, Silman I (1991) Atomic structure of acetylcholinesterase from torpedo californica: a prototypic acetylcholine binding protein. Science 253:872–879

Stubbs MT, Oschkinat H, Mayr I, Huber R, Angliker H, Stone SR, Bode W (1992) The interaction of thrombin with fibrinogen – a structural basis for its specificity. Eur J Biochem 206:187–195

Von Itzstein M, Wu W-Y, Kok GB, Pegg MS, Dyason JC, Jin B, Van Phan T, Smythe ML, White HF, Oliver SW, Colman P, Varghese JN, Ryan DM, Woods JM, Bethell RC, Hotham VJ, Cameron JM, Penn CR (1993) Rational design of potent sialidase-based inhibitors of influenza virus replication. Nature 363:418–423

Winkler FK, D'Arcy A, Hunziker W (1990) Structure of human pancreatic lipase. Nature 343:771–774

Winkler FK, Banner DW, Oefner C, Tsernoglou D, Brown RS, Heathman SP, Bryan RK, Martin PD, Petratos K, Wilson KS (1993) The crystal structure of EcoRV endonuclease and of its complexes with cognate and non-cognate DNA fragments. EMBO J 12:1278-1795

5 Chances and Risks of Modeling in Industry

H. Köppen

5.1 Introduction

After more than a decade of development, molecular modeling is an established technique in the pharmaceutical industry. It started in the late 1970s/early 1980s basically as a tool to visualize molecular shape but it soon became obvious that computer-based modeling offers more than only graphics: the still ongoing improvement of force fields allows optimization of the geometry of molecules as well as assessment of their flexibility. Quantum mechanical calculations yield both minimum energy conformations and electron densities, for example – if required at a high level of accuracy. Both the storage capacity and the computational speed of modern hardware assist in comparing the molecular properties of many interesting molecules in a systematic way which goes far beyond the limits of mechanical models. The key features which distinguish active and inactive molecules can be deduced from that comparison. The interested reader should refer to some of the reviews in this field for more details (Cohen et al. 1990; Kubinyi 1993; Lipkowitz and Boyd 1994).

However, the success of modeling applications in industry depends on more than good scientific practice, the availability of appropriate algorithms, and access to modern hardware and software. The integration of modeling groups into biological and chemical research and the communication between computational chemists and other research team members is also crucial. The industrial environment offers many chances but may also bear some risk for computer-aided drug discovery applications and for modeling groups.

This paper focuses on these communication and integration aspects. It also briefly describes some of the advances which have been made during recent years, some of the remaining deficiencies, and the impact of computational techniques on screening approaches.

5.2 Are New Drugs Designed or Discovered?

It is important to note that current applications of computational chemistry encompass not only all kinds of conventional drug optimization but also "de novo design" of ligands (Böhm 1993; Blaney and Dixon 1993; Moon and Howe 1991; Kuntz 1992) and issues related to screening (Hodes 1990; Craig 1990; Lajiness 1991). In contrast to former times, computational tools play a role even in the very first stages of research for a new drug. The term "computer-aided drug discovery" is therefore preferred over the term "molecular modeling" which is too narrow a description of today's applications. It is also preferred over the term "computer-aided drug design" since a true "design" of new drugs is not yet feasible. Too many parameters relevant for both pharmacokinetics and pharmacodynamics of a compound are not yet known. Still, new drugs are more often discovered than designed even if computational techniques play an increasingly important role for the efficient development of a compound from screening to clinics.

Computer-aided drug discovery is no replacement for synthesis and biological testing. It was a common misconception in the early years to expect that new drugs could be "designed" in the same way as a new car or an aircraft is. While in such cases nearly all of the underlying science and technology is known and the problem is basically to optimize the design and production process, there is much information missing regarding the interaction of a new compound with an enzyme or receptor. Hence, computer-aided drug discovery does not provide a straightforward way to success of pharmaceutical R&D by eliminating the risks of dealing with new compounds and new biological targets. At times, however, computational chemists might have promoted such a view in overselling computer-aided drug discovery to a skeptical management in order to get the necessary investments approved. On the other hand, one may imagine as well that some management boards faced with the threat of rapidly increasing R&D costs and times were only too willing to believe that computer-aided drug discovery would be the panacea producing the "big score" if first-rate graphics and supercomputers are provided. It is no surprise that these unrealistic expectations on what computer-aided drug discovery can contribute led to disappointment.

Those who are responsible for the budget and for the investments of a modeling group should carefully avoid raising unrealistic expectations. Nevertheless, they have to make sure that their group has access to state-of-the-art technology. This is a difficult balance. Computer-aided drug discovery is a discipline which relies on cutting-edge hardware and software. The progress made in computer-aided drug discovery during the last 10–15 years would be unthinkable without the progress made in hardware development. This does not mean at all that powerful computers or high resolution graphics alone are sufficient to guarantee success. The development of sophisticated algorithms as well as access to a rapidly growing number of experimental data (e.g., X-ray structures of proteins) is absolutely essential for computer-aided drug discovery. But without modern hardware both these software tools and the huge amount of experimental data could not be used efficiently. The handling of three-dimensional (3D) databases or the replacement of the often inadequate semiempirical approaches by higher level ab initio calculations, to name only two examples, require fast machines. More powerful hardware enables the scientists to use more sophisticated models and to improve the quality of the results – an aspect which is more important than merely increasing the throughput. It is no trivial task to make sure that the management recognizes well this aspect of quality improvement without getting the impression that former contributions from computer-aided drug discovery were scientifically doubtful. It is also important to point out to the management that, in contrast to many commercial applications, more powerful hardware for computer-aided drug discovery does not mean that personnel can be reduced.

5.3 Aspects of Interdisciplinary Work

A characteristic feature of the industrial environment is the close integration of a broad range of scientific disciplines into the pharmaceutical research and development process. The scientists are typically members of an interdepartmental research team. It is most important for the success of computer-aided drug discovery that the computational chemist not only provides service on request for such a team but that he or she be a team member as well. In this way a frequent and intensive exchange of experimental facts and new ideas can take place and stimulate the research process, which is always a cycle of hypothesis generation, testing, and refinement.

In contrast, there are many academic groups who have developed excellent scientific theories but who have serious problems to getting a molecule which they have suggested as an interesting new ligand for a biological target synthesized and tested. They also have often problems getting consistent biological data and carefully designed series of molecules from the literature in order to test a new approach or algorithm. In Germany, but presumably also in other European countries, close collaboration between different scientific disciplines in the academic environment is still more the exception than the rule. This may also be a problem in industry from time to time but, in general, the computational chemist may get interesting new compounds synthesized and tested. Computational chemists have access to all the necessary information on the details of biological tests as well as to the large number of proprietary structural and biological information, which allows them to develop structure–activity hypotheses. It is worth mentioning in this context that a complete (in terms of data) and user-friendly (in terms of easy handling and short response times) in-house research documentation system is an indispensable tool for computer-aided drug discovery.

5.4 Limitations of Computer-Aided Drug Discovery

Computer-aided drug discovery provides tools to generate hypotheses about structure–activity relationships. It complements our information concerning structures and biological data by adding physical principles in terms of force fields, quantum mechanics, and thermodynamics. It also makes it possible to consider large series of molecules, taking advantage of the storage and computational facilities of modern hardware. If done properly hypotheses generated in this way (e.g., pharmacophore models) are the best possible representation of our current knowledge. However, two limitations must be kept in mind:

Complex influences such as entropy changes, solvation, and desolvation or long-range electrostatic interactions are not adequately taken into consideration in most cases. Calculated energies are typically only enthalpies and not free enthalpies (Abagyan 1993; Andrews 1993). The huge computational efforts required by more precise models may lead the theoretician to prefer less accurate approaches but there are also rather trivial reasons for limited precision, for example, missing parameters needed for the fine tuning of a force field (a very common case). Therefore the accuracy of a model may be reduced because of limitations in computer-aided drug discovery techniques. Some of these limitations may be overcome by using powerful hardware but the appropriate treatment of others goes far beyond the current limits of simulation techniques (e.g., long-range conformational changes of proteins such as the folding problem or diffusion processes).

Another problem may arise from our limited knowledge of the biological system which we are dealing with. This is obvious with respect to the complex path of a drug through the whole living body. But even test systems which seem to be much simpler may yield puzzling data. Isolated organs still contain a variety of living cells with a whole battery of receptors and ion channels. These may interact in a complex, not yet well understood way. It is not always clear how to interpret a biological signal. Is it the interaction of a compound with only one target or the superposition of several biological events? How do on- and off-kinetics influence the results? What role does the interaction with the cell membranes play? Are the data time dependent because of up- or down-regulation of receptors? And not to forget: Does the compound precipitate unexpectedly when transferred from the stock solution into the liquid used for the biological test (which may be blood, for example)? Or is it partly adsorbed at the walls of the test system? The concentration of a solute may change without being noticed, which leads to wrong experimental results. Even if we have a pretty isolated target, e.g., an enzyme or a well-defined receptor subtype, we may be faced with the so-called "multiple binding mode" problem, where ligands of more or less similar shape bind in surprisingly different ways to a given binding site (Mattos and Ringe 1993). Obviously, pharmacophore models derived by superposition of these ligands are as useful as a comparison between apples and oranges. There is no doubt that X-ray data should be used whenever possible in order to avoid such misleading conclusions. Unfortunately, there are cases where X-ray data are not yet available, especially for membrane-bound proteins. It is, however, encouraging to see that new techniques such as atomic force microscopy are emerging which may fill that gap (Frisbie et al. 1994; Radmacher et al. 1994).

5.5 Communication Aspects

All these limitations must be discussed in an open atmosphere within the research team. This requires open-minded scientists willing to critically review their own results as well as accepting questions from other team members regarding the meaning of the presented results. This in-depth discussion of results caused by questions arising from an unexpected point of view and asked by team members from other disciplines offers the chance to put the data into a new context and may lead to a deeper understanding of the experimental results. However, these questions may also be misunderstood as criticism with respect to the scientist's qualification, in particular if it concerns a discussion between different disciplines. The respective scientist may tend to consider the other team members as being not qualified enough to express any criticism, which, in turn, creates frustration and continuous communication problems within the research team. In addition to their scientific qualification, members of an interdisciplinary team therefore need a large amount of social competence and communication skills.

This is even more important for computational chemists, who are often regarded as theoreticians whose contributions to the project's progress are not always as obvious as the contributions of those who provide "real" compounds or experimental data with impact, for example, for drug registration. This skepticism may frustrate the computational chemist and stimulate him to put even more effort into a rigorous theoretical treatment of the structure–activity relationships in order to demonstrate the value of computer-aided drug discovery. This will take some time, which is used by the other team members to go ahead with research, without taking much notice of the computational chemist's work. In this way computer-aided drug discovery will always be behind the current status of research. This is a most serious problem: If computer-aided drug discovery is not able to provide timely solutions, the other scientists will basically return to traditional techniques, such as 2D superpositions.

Pharmaceutical research using these traditional techniques has given us nearly all the drugs which we use today. They have definitely been successful. This fact is turned into one of the most often heard arguments raised by skeptical scientists against computer-aided drug discovery. But it is not a question of whether this kind of research was successful in the past, but rather whether companies will survive on a long-term scale if they do not systematically and intensively use all available approaches including computer-aided drug discovery to optimize the effectiveness of their research process. The rapidly emerging techniques of combinatorial libraries and robotic synthesis combined with high-throughput screening will change traditional pharmaceutical research in the near future. Only those companies which are able to analyze the huge amount of structural and biological data, identify the subtle structure–activity patterns, and use this information effectively for the optimization process will remain competitive.

These considerations, however, do not help the research team if the modeling group is not able to keep pace with the progress of the team. Faster hardware may be supportive in some cases but it is more important for the computational chemists to focus on those aspects of the research project where adequate computational techniques are available, where support is most needed, and where the chance to get proposed compounds synthesized is most realistic. This aim can be achieved only if there is continuous, close collaboration between medicinal chemists, biologists, and computational chemists.

5.6 The Integration of Computer-Aided Drug Discovery and Medicinal Chemistry

First of all, computer-aided drug discovery provides only figures and pictures – no compounds. It may generate excellent hypotheses using cutting-edge scientific and technological tools, but without experimental feedback computer-aided drug discovery will not be of any benefit for research projects and, hence, remain only a high-tech way of wasting money. Therefore, computer-aided drug discovery must be tied into synthesis and biological testing. Modeling groups do not synthesize molecules and they can only indirectly influence research success. The management should therefore aim at establishing the appropriate organizational structures which facilitate the interaction between computer-aided drug discovery and experimentalists.

Any organizational scheme may work if people make it work. However, departmental boundaries may interfere with the required close collaboration between experimentalists and theoreticians. A charge-back system, for example, may be implemented which ensures that scientists from other departments who are getting support from the modeling group share some amount of the financial burden caused by this group. There is a high risk that such a charge-back system may frighten off those who are interested in using computer-aided drug discovery techniques in particular since it cannot always be well predicted how much effort and costs a given problem will require. It is even harder to predict how much the computer-aided drug discovery support will contribute in terms of the overall project progress. In that respect computer-aided drug discovery should be considered as a "strategic company resource" similar to the libraries, for example, in order to encourage intensive use. It can easily be envisaged how the use of literature data would be reduced if every scientist entering the library had to pay a significant fee, depending on the number of books or journals that are going to be read. It can also be envisaged how such a restrictive access to information would impair the company's long-term chances to survive. Computer-aided drug discovery also provides information about structure–activity relationship hypotheses. This information helps optimize the chances for success of research projects, but success is not guaranteed, as it is typical of all innovative research projects in pharmaceutical industry.

We found the integration of computer-aided drug discovery into the medicinal chemistry department to be the best solution. The medicinal chemists are those scientists whose collaboration is most essential for the success of computer-aided drug discovery. Having both groups in one department enhances their mutual contacts and encourages the scientists to get in touch with each other. It also avoids fruitless discussions about a charge-back system since the budget distribution within a department is typically easier than between different departments. Moreover, the department head has influence on capacities and can direct the computer-aided drug discovery's priorities. There is always some risk that a small group of theoreticians will tend to live in an ivory tower, optimizing basically their own tools. Within the medicinal chemistry department there are many ways to integrate computer-aided drug discovery into all discussions, ranging from the biology of new projects to special problems of synthesis. In this way it can be made sure that modeling group members do not drift away from real life. Computational chemists sometimes tend to suggest new structures, for example, derived from de novo design programs, whose synthesis is, in fact, challenging even to very experienced organic chemists. Modelers in general prefer rigid compounds such as spirosystems which are often difficult to synthesize. It is highly recommended that the necessary efforts put into synthesis be balanced against the quality of the hypothesis! To suggest a time-consuming, 15-step synthesis based on a crude model is very risky and should be avoided. Should the prediction fail, the chemist may have problems justifying the efforts. He may then be tempted to suggest to the management that computer-aided drug discovery gave absolutely misleading hints. It is a great advantage of in-house 3D databases that the compounds are either still available or that at least the synthetic routes to all molecules are known.

Our group is involved in many projects discussions in an early stage even if work on the actual synthesis has not yet begun. The assignment of computer-aided drug discovery to projects in an early stage is important since it may turn out that the specific requirements of the upcoming project require some basic research in terms of, for example, force field tuning or validation of quantum mechanical approaches. These evaluation studies require some time. If the modeling group is going to be involved in later stages of the project this time is missing. The other research team members expect, however, that the computational chemists are well prepared to assist them at any time. Such a situation typically leads to the already mentioned problems of computer-aided drug discovery not being able to keep pace with research. Even in such an unpleasant situation, it cannot be recommended that thorough scientific work be replaced with quick and dirty solutions. There are cases where even crude models help to decide what should be synthesized, measured, or computed next. But done carelessly, such an approach may seriously damage the reputation of computer-aided drug discovery. Instead, one should discuss frankly the situation even if it turns out that the only solution is to refrain from using computer-aided drug discovery for that project.

The relationship between computer-aided drug discovery and medicinal chemistry deserves some special remarks. Over many decades the chemists were the only ones who invented drugs. Any new structure was planned first on a sheet of paper in their office with a respective priority on the patent claim. This kind of single-combat mentality where each scientist is competing with all others has no future anymore and is being replaced more and more by a team approach. In this respect the input from modeling must also be considered as part of this team approach. It is essential that the management not only count the number of newly synthesized compounds of individual chemists but that the whole team is awarded for a success. Otherwise the chemists might see the modelers as competitors and try to demonstrate that computer-aided drug discovery support is neither needed nor useful at all instead of making best use of suggestions from modeling.

In order to avoid any competition between computer-aided drug discovery and medicinal chemists we have tried to establish that the chemists use modeling themselves. Nearly all laboratory heads got training in basic modeling techniques and free access to the screens which were located in close vicinity to the labs. Computational chemists were available to assist them. However, we realized that this method of integrating medicinal chemistry and modeling does not work well.

Computer-aided drug discovery is a full-time job which needs a lot of continuous training and special scientific background. To be in charge of heading a lab and doing intensive modeling studies at the same time is a conflict which cannot be resolved in most cases. Today, we still offer access to the workstations but this offer is rarely taken up. Instead, the chemists typically go to the modeler and discuss structure–activity relationships and ideas about new structures with him in front of the screen. It is important that the distance between the chemist's labs and computer-aided drug discovery's offices not be too great! If possible, the groups should be located in the same building in order to keep the activation barrier for frequent visits low. Despite the fact that modeling in our company is meanwhile nearly exclusively done by specialists we found that the training of chemists in computer-aided drug discovery techniques is useful in order to reduce "language problems" between them and the modelers.

The availability of powerful, networked PCs on each chemist's desk may change this situation in the future. We do not expect that chemists will do all the modeling work themselves again but rather that they will use their PC as a user-friendly window to have a look at the results of the work done by the modeling group. We expect that this will enhance the communication between these two groups.

5.7 Relation Between Computer-Aided Drug Discovery and Management Information Systems

The basis of computer-aided drug discovery is, of course, information processing. Both the selection of new hardware as well as the technical support of the considerable computing power which is typically accumulated in a modeling group may be an issue between computer-aided drug discovery and the management information systems (MIS) department. Computer-aided drug discovery needs very powerful hardware, ranging from high-performance workstations or even true supercomputers to high-resolution real-time graphics, hardware at the cutting edge of development which therefore cannot have been established for years already. Dealing with such hardware and with the respective operating system and application software requires both special training and an increased resistance to frustration by sudden hardware breakdowns or software bugs. This is no excuse for vendors to sell underdeveloped products but it is in complete contrast to the general strategy of a central MIS department, which for good reasons prefers well established hardware and software. The typical MIS customers are secretaries or bookkeepers who hardly need supercomputing power but rather very reliable and user-friendly technology. Hence, the MIS department is much more familiar with these requirements and sometimes reluctant with respect to the special requests from computer-aided drug discovery.

Our experience is that decisions about hardware and software should be made by the modeling group after intensive discussions with the MIS department. It is recommended that aspects of the MIS department be taken seriously into account since computer-aided drug discovery cannot be an isolated island in a complex information technology environment. Workstations are networked, support from UNIX specialists may be needed, and the modeling group members may need to access central resources such as file or print servers. It is therefore much better to establish smooth cooperation between computer-aided drug discovery and MIS. The MIS department, on the other hand, should accept the technological expertise of a modeling group in this special field.

In our group, a technician maintains the dedicated computer-aided drug discovery hardware and software. In this way it is ensured that the necessary expertise is available on request. However, MIS service is used whenever possible in order to keep the number of local personnel down.

5.8 Capacity Considerations

The personnel capacity needed for modeling may be a serious problem. Computer-aided drug discovery requires special training as well as scientific background, and computer-aided drug discovery projects are often involved and time consuming. Since all pharmaceutical companies are faced with cost reduction programs personnel capacity in particular is an issue. We have found that a full-time computer-aided drug discovery scientist cannot handle more than two projects at the same time. If there are additional responsibilities, overhead because of administration, membership in internal or external working groups etc. this number is even reduced to one.

It is our experience that the number of projects in which computer-aided drug discovery support can be useful always exceeds the capacity of a modeling group. The decision as to which projects should be supported by computer-aided drug discovery is difficult to make, but most important. A restrictive attitude means refusing the requests of some medicinal chemists for support; they may be hesitant to ask again even if they have a project which is well suited for computer-aided drug discovery approaches. The "friendly modeler" attitude, on the other hand, may lead to many simultaneous projects with the risk of long response times and/or lack of scientific quality because of capacity problems. This may cause disappointment as well.

It is important to discuss these aspects with the chemists. It may also be very helpful if they have some background in modeling techniques. In this way unrealistic expectations are at least reduced and the chemists get a better feeling for the required computer-aided drug discovery capacity. Projects where 3D data about the target are available are to be preferred of course. It is always recommended to have a close look at the biological data. Sometimes compounds which are said to be "more active" in fact bind with lower affinity to the biological target. Only their bioavailability is higher. Other compounds which are said to be "inactive" do indeed strongly interact with the receptor as long as they are not metabolized. It is recommended that the biological data be discussed carefully before starting a modeling project. Puzzling biological data should be considered with care.

5.9 Progress of Computer-Aided Drug Discovery and Still Existing Limitations

It is not the aim of this chapter to discuss in detail the scientific progress during the last years or the remaining deficiencies of modeling and computer-aided drug discovery in general. Instead, only a few aspects should be highlighted which we consider as most significant.

Both the de novo design techniques and the 3D databases represent a very significant advancement. De novo design programs make it possible to suggest completely new molecules to the chemists. They can be used even in cases where only a pharmacophore model is available, but their real value is the docking of de novo-designed ligands into the binding site of a target with known 3D structure (Böhm 1993; Blaney and Dixon 1993; Moon and Howe 1991; Kuntz 1992). In answering the question "what should be synthesized next," they go beyond the limits of traditional modeling techniques which can only manage already known structures. Not all suggestions, however, are reasonable from the point of view of synthesis and it would be desirable to complement de novo design programs by modules which check the synthetic feasibility. In any case results from de novo design are extremely stimulating even if the compound suggested by the program must be modified from the chemist's point of view.

3D databases are a kind of powerful machinery designed to perform geometrical and conformational analysis on extremely large sets of compounds (Martin 1992; Pearlman 1993) – a task which was absolutely impossible in the past. A specific 3D query based on a pharmacophore model is used as input. The hit list consists of compounds which fulfill the geometrical criteria of the query, taking into account the conformational flexibility of the structures stored in the database. The hits may be put into a biological test or may serve as a suggestion for further synthesis. Another application is to check whether the pharmacophore model is consistent with all available data since active compounds should be found as hits while inactive compounds must not be found if the 3D query is specified in sufficient detail. In any case outliers can be identified, analyzed separately, and be used to refine the model.

As described 3D databases may help to identify compounds which may be available on stock, may fit to the receptor, but have not yet been tested. This is a breakthrough for computer-aided drug discovery since the test of available compounds is the fastest experimental feedback which modeling can get at all. On the other hand, one has to be cautious in interpreting negative results: It may happen that a selected compound has all features which are assumed to be essential for receptor binding. It is therefore found among the hits. However, some parts of the molecule which are not considered in the query may interfere with the spatial or electronic requirements of the receptor. This inactive compound could be converted into an active one by minor structural modifications. This may be a problem but also a strength of that approach. It is sometimes argued that high-throughput screening would be able to identify all active compounds available in a company's dispensary within a short time. Hence, there would be no need for 3D databases of in-house compounds. This is certainly not true with respect to the above-mentioned consistency check of pharmacophore models. But it is also not true with respect to all compounds which are inactive only because of slight structural deviations from the optimum. They cannot be found in any screening but there is a chance of identifying them using a suitable 3D query.

Among the still-existing limitations of computer-aided drug discovery aspects of bioavailability are most significant. There are many well-developed tools today to deal with aspects of ligand–receptor interaction, such as conformational or shape analysis of ligands, docking, or comparison of electrostatic fields. However, the solubility of compounds, the resorption, the penetration of skin or blood–brain barrier or the metabolism are less well understood and cannot yet be predicted in many cases. This is a serious limitation. It is our experience that in many projects there are enough ligands with high receptor affinity. It is basically the pharmacokinetic behavior of the compounds which makes them unsuitable as drugs. It would be highly desirable to have molecular parameters at hand which allow the blood–brain distribution to be predicted. There are some publications dealing with these issues but we are still far away from any breakthrough (Bassolino-Klimas et al. 1993; Seelig et al. 1994; Chikhale et al. 1994; Abraham et al. 1994).

We have found that medicinal chemists sometimes complain that modeling may be useful for designing a high-affinity ligand but computer-aided drug discovery does not really help in getting that ligand to the receptor. This is a serious obstacle for further development. The scientific development of computer-aided drug discovery should therefore focus more on these aspects. Progress in that direction will be highly appreciated by all research teams.

5.10 Impact of Computer-Aided Drug Design on Screening

It is remarkable that screening, which in the past was considered as a completely random approach, is now taking more and more advantage of computer-aided drug discovery techniques. There are various aspects to be mentioned:

- All pharmaceutical companies have large, high-throughput screening assays running. They all have collections of in-house compounds on stock too. These compounds represent a tremendous commercial value since the synthesis of a single compound typically causes in-house costs of some thousand DM.
- Compounds which are not available anymore will hardly ever be resynthesized. It is therefore necessary to optimize the use of these compounds in screening assays. We are going to cluster our in-house compounds according to their 2D similarity using the Daylight clustering package (Daylight 1994). Representatives of each cluster are tested first. This approach is not intended to exclude compounds from screening but rather to set up a priority list based on dissimilarity. We expect that the hit rate may be increased in this way and screening can be stopped earlier. Up to now there are only few data available in the literature, but they do support this hypothesis (Lajiness 1991; Hodes 1990).
- Another aspect is the selection of compounds from commercial suppliers. These compounds should represent new structural prototypes in order to complement the in-house database. Using the Daylight software, again, we found that in a typical case less than 10% of the external compounds represented new structural prototypes (H. Briem 1994, unpublished results) different from those in our in-house database. Purchasing only these prototypes may be the most economical way of complementing the in-house database.
- A new and promising technique in the context of screening is the design of compound libraries (Hobbs DeWitt et al. 1993; Burgess et al. 1994; Zuckermann et al. 1994; Gallop et al. 1994). This technique will drastically change pharmaceutical research, in particular, if versatile organic compound libraries with millions of different compounds can be developed. On the other hand, this figure clearly shows that design criteria must be used in order to cover the space of physicochemical parameters with a reasonable number of compounds. Some of the traditional quantitative structure–activitiy relationships (QSAR) techniques may experience a kind of revival in that context. Another problem will be to analyze the huge amount of screening data with low signal-to-noise ratio in order to identify subtle structure-activity patterns. This field is still under development and standard solutions in terms of algorithms and software

products are not yet established. It is obvious, though, that computer-aided drug discovery techniques will be needed to manage that task. It can be envisaged that those companies which are able to handle these data problems in the most efficient way will be most competitive.

5.11 Conclusions

Computer-aided drug discovery is an established technique used by nearly all pharmaceutical companies in order to optimize their research and development. Today computer-aided drug discovery has a much broader scope than was expected 10 years ago, also covering aspects of screening support and lead compound design. It is encouraging to see that though faced with tough cost reduction programs hardly any company is going to reduce the personnel capacity of modeling groups.

On the other hand, there is still much room for improvement, in particular with respect to the optimization of pharmacokinetic properties. Progress towards this end would be highly appreciated.

Computational chemists provide tools and knowledge to generate hypotheses concerning structure–activity relationships. Since experimental feedback is indispensable collaboration with many other scientists is required in order to get compounds synthesized and tested. Only if the interdisciplinary communication aspects can be managed in addition to all the scientific issues can and will computer-aided drug discovery play an important role in the development of a new drug from screening to clinical application.

References

Abagyan RA (1993) Towards protein folding by global energy optimization. FEBS Lett 325:17–22

Abraham MH, Chadha HS, Mitchell RC (1994) Hydrogen bonding 33. Factors that influence the distribution of solutes between blood and Brain J Pharm Sci 83:1257–1268

Andrews PR (1993) Drug-receptor interactions. In: Kubinyi H (ed) 3D QSAR in drug design. Escom, Leiden, pp 13–40

Bassolino-Klimas D, Alper HE, Stouch TR (1993) Solute diffusion in lipid bilayer membranes: an atomic level study by molecular dynamics simulation. Biochemistry 32:12624–12637

Blaney JM, Dixon JS (1993) A good ligand is hard to find: automated docking methods. Perspect Drug Discov Design 1:301–319

Böhm H-J (1993) LUDI: rule-based automatic design of new substituents for enzyme inhibitor leads. J Comput Aided Mol Design 6:593–606

Burgess K, Liaw AI, Wang N (1994) Combinatorial technologies involving reiterative division/coupling/recombination: statistical considerations. J Med Chem 37:2985–2987

Chikhale EG, NG K-Y, Burton PS, Borchardt RT (1994) Hydrogen bonding potential as a determinant of the in vitro and in situ blood-brain barrier permeability of peptides. Pharm Res 11:412–419

Cohen NC, Blaney M, Humblet C, Gund P, Barry DC (1990) Molecular modeling software and methods for medicinal chemistry. J Med Chem 33:883–894

Craig PN (1990) Substructural analysis and compound selection. In: Ramsden CA, Sammes PG, Taylor JB (eds) Comprehensive medicinal chemistry, vol 4. Pergamon, Oxford, pp 645–666

Daylight clustering package (1994) Version 4.3. Daylight Chemical Information Systems, Irvine, USA

Frisbie CD, Rozsnyai LF, Noy A, Wrighton MS, Lieber CM (1994) Functional group imaging by chemical force microscopy. Science 265:2071–2074

Gallop MA, Barrett RW, Dower WJ, Fodor SPA, Gordon EM (1994) Applications of combinatorial technologies to drug discovery. 1. Background and peptide combinatorial libraries. J Med Chem 37:1233–1251

Hobbs DeWitt S, Kiely JS, Stankovic CJ, Schroeder CM, Reynolds Cody DM, Pavia MR (1993) "Diversomers": an approach to nonpeptide, nonoligomeric chemical diversity. Proc Natl Acad Sci USA 90:6909–6913

Hodes L (1990) Computer-aided selection for large-scale screening. In: Kennewell PD, Sammes PG, Taylor JB (eds) Comprehensive medicinal chemistry, vol 1. Pergamon, Oxford, pp 279–284

Kubinyi H (ed) (1993) 3D QSAR in drug design – theory methods and applications. Escom, Leiden

Kuntz ID (1992) Structure-based strategies for drug design and discovery. Science 257:1078–1082

Lajiness MS (1991) An evaluation of the performance of dissimilarity selection. In: Silipo C, Vittoria A (eds) QSAR: rational approaches to the design of bioactive compounds. Elsevier, Amsterdam, pp 201–204

Lipkowitz KB, Boyd DB (eds) (1994) Reviews in computational chemistry. VCH, New York

Martin YC (1992) 3D database searching in drug design. J Med Chem 35:2145–2154

Mattos C, Ringe D (1993) Multiple binding modes. In: Kubinyi H (ed) 3D QSAR in drug design. Escom, Leiden, pp 226–254

Moon JB, Howe WJ (1991) Computer design of bioactive molecules: a method for receptor-based de novo ligand design. Protein Struct Funct Genet 11:314–328

Pearlman RS (1993) 3D molecular structures: generation and use in 3D searching. In: Kubinyi H (ed) 3D QSAR in drug design. Escom, Leiden, pp 41–79

Radmacher M, Fritz M, Hansma HG, Hansma PK (1994) Direct observation of enzyme activity with the atomic force microscope. Science 265:1577–1579

Seelig A, Gottschlich R, Devant RM (1994) A method to determine the ability of drugs to diffuse through the blood-brain barrier. Proc Natl Acad Sci USA 91:68–72

Zuckermann RN, Martin EJ, Spellmeyer DC, Stauber GB, Shoemaker KR, Kerr JM, Figliozzi GM, Goff DA, Siani MA, Simon RJ, Banville SC, Brown EG, Wang L, Richter LS, Moos WH (1994) Discovery of nanomolar ligands for 7-transmembrane G-protein-coupled receptors from a diverse N-(substituted)glycine peptoid library. J Med Chem 37:2678–2685

6 The Advantages of Using Rational Drug Design in Modern Drug Discovery: How to Integrate Computer-Aided Drug Design and Modern Biotechnology

U. Norinder

6.1 Introduction

The research efforts invested in the development of a new drug are considerable today. It may take more than 15 years from the start of research to the approval of a new product. During that time a vast number of compounds have been synthesized and tested for every viable product. For a long time now the pharmaceutical industry has been interested in finding new techniques that will reduce the number of substances in the early stages of development and identify entities with a high probability of success through the clinical phases to attain approval as a new drug.

Today biotechnology is one of the most rapidly advancing scientific disciplines. It is also one of the most important research areas for the pharmaceutical industry. Biotechnology will probably have the greatest impact in the discovery of new drugs. Most biotechnology-related products presently on the market are protein or peptide based and cannot be orally administered. The knowledge gained by biotechnological research will generate the next generation of drugs, among which will be orally acting, small peptidomimetic compounds (organic molecules).

One of the major objectives of the pharmaceutical industry, as touched upon, is to bring new and effective drugs and other products to the market quickly and cost-effectively. This, apart from chance or sheer luck, requires efficency both as far as organization and research instruments, e.g., methodologies and techniques such as computer-aided drug design (CADD), are concerned.

6.2 Drug Discovery Using CADD

There are some conditions which facilitate the integration of CADD as a useful part of drug discovery. Among them is close interdisciplinary contact within the organization. In small companies, both with respect to the number of employees and size of the premises, this is usually not too troublesome to achieve. Another point of interest is the pace within the testing–model building–organic synthesis cycle (Table 1).

The test systems used should be rapid and reproducible and give relevant information about the tested entities for the target in question. Otherwise, the computer-aided drug discovery process will easily lose

Table 1. Conditions which facilitate the integration of CADD in the drug discovery process

1. Interdisciplinary contact
2. Size and premises of the organization
3. Pace of the testing–model building–organic chemistry cycle

Table 2. Test systems

1. Rapid
2. Reproducible
3. Relevant information for the target
4. Automation/robotics
5. Laboratory information management systems

Table 3. Biotech systems

1. Satisfy points 1–4 in Table 2
2. Postpone animal studies
3. Place too much faith in biotechnological test methods
4. Sometimes difficult to develop
5. Ensure a higher probability of relevant compounds for animal testing

Table 4. Models

1. Structure–activity models (general direction)
2. Balance between the use of models and knowhow of chemists
3. Level of understanding from models
   Pragmatic
   Mechanistic

vitality and momentum and even come to a halt where its use is doubted and/or severely questioned (Table 2).

In this respect, biotechnologically based test methods are good. Most of them meet the criteria mentioned above , which make them particularly useful in drug research. Furthermore, these test methods may postpone animal studies until a later stage in drug development. They may also ensure there is a higher probability that the tested compounds will have a more favorable profile once they reach the animal testing stage. Exchanging some aminal studies for in vitro test systems is also beneficial with respect to the complex of problems when using animals for testing. Not having to rely on animal-based test models until later stages in the drug development process saves considerable time and money and makes it easier to keep a high pace in research projects (Table 3). One disadvantage, as with all methods, is of course that one may rely too much on the results from the biotechnological methods and miss compounds that would have proven valuable by other methods for further development.

As to the development and use of structure–activity models there should be a balance between their use and the professional skills of the scientists (chemists) involved in synthesis and testing of new compounds. The models should indicate a general direction or trend to pursue. However, they should not be used in a dogmatic manner but rather encourage initiatives to be taken along this path. One may divide the quantitative model typess into two categories: those that are pragmatic in nature, i.e., gray or black boxes, which means that they are difficult to interpret but good at predicting new, interesting molecules to synthesize; and those models which aid in the understanding of what properties and structural entities are important for the target in question and even help in elucidating the underlying mechanisms of action/interaction. The former models are certainly easier to derive while the latter ones usually require considerably more time to develop. However, research projects in the pharmaceutical industry do not always allow for the latter models to be developed since a primary objective in pharmaceutical research is to develop entities that result in a new drug on the market in as short time as possible in a cost-effective manner (Table 4).

Karo Bio AB is a biotechnology company engaged in drug discovery using modern technologies, which involves a combination of rational drug design incorporating CADD and experimental design as well as synthetic organic chemistry, coupled with rational screening techniques based on engineered cell lines that are stable and reproducible and which permit a rapid throughput of test compounds to achieve the objective of effectiveness (Fig. 1).

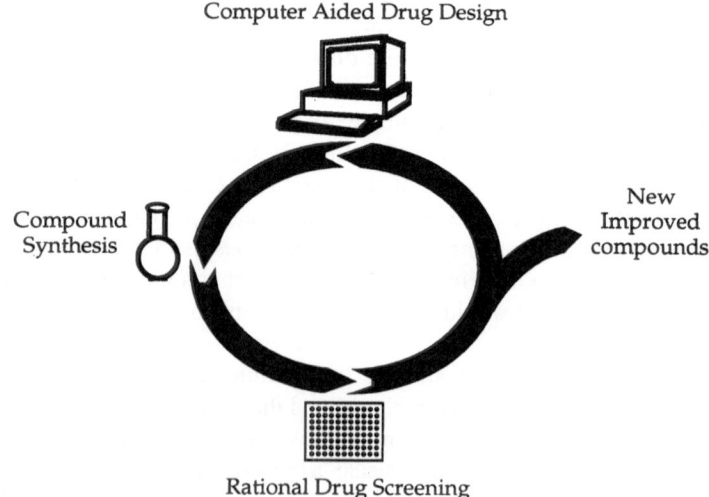

Fig. 1. Karo Bio drug discovery system

6.3 Rational Drug Screening

The drug screening system is in vitro based and determines the functional consequence of a substance interacting with a target receptor. The system consists of a primary, secondary, and tertiary subsystem (Fig. 2).

The primary screening system uses genetically engineered cells which make it possible to determine agonistic or antagonistic effects (Table 5). The secondary screening system is based on the availability of receptors of high quality. With this system the binding affinity of a

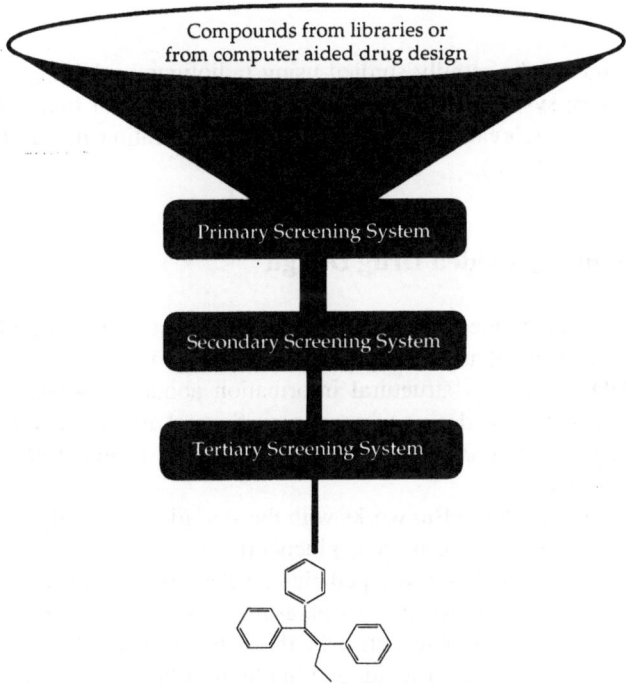

Identification of tissue selective receptor
agonists/antagonists for testing in
relevant animal models

Fig. 2. Rational drug screening

Table 5. Reporter system stage 1 cell lines

1. Stable cell lines
2. Agonist/antagonist activity
3. Potency of compounds
4. Specificity of compounds

Table 6. Reporter system stage 2 cell lines

1. Custom designed
2. Selectivity in cell lines derived from different tissues
3. Agonist/antagonist activity on endogenous markers
4. Value for the indication in question

compound can be rapidly studied using radioligand inhibition assays. The tertiary systems have nonmanipulated cell lines of human origin derived from relevant target tissues for the indication in question as their basis (Table 6).

6.4 Computer-Aided Drug Design

CADD can be divideed into two main parts: direct (or structure-based) drug design (DDD) and indirect drug design (IDD).

In DDD one uses structural information about a macromolecular target to assist in the design of new drugs. Several articles have recently discussed this topic (Greer et al. 1994; Navia and Peattie 1993; Reich and Webber 1993).

The company Karo Bio works with the steroid superfamily of receptors, in particular the estrogen, glucocorticoid, and thyroid hormone receptors. Karo Bio has developed high level expression and production systems of the target proteins as well as assays to monitor the quality of the produced proteins. Currently, the three-dimensional (3D) structures of the estrogen, glucocorticoid, and the thyroid hormone receptors are being determined using X-ray and nuclear magnetic resonance (NMR) techniques as well as other spectroscopic methods. The goal of these efforts is to extract knowledge not only of the ligand-binding domain and of the protein but also to obtain information regarding the dynamics of the system. The latter will provide valuable information about gene activation and tissue selectivity.

Atomic-level information about the 3D geometry of the active site of the protein will most probably facilitate the discovery of new lead compounds. By using this knowledge one may identify new amino acid residues or areas thereof capable of interacting with a potential new lead structure. One also obtains valuable information as to the spacial restraints of the active site.

However, in most cases it is not a trivial task to establish a direct quantitative correlation between the points of interaction of the protein with the ligand molecule. Molecular dynamics calculations at a sufficient level of approximation is still a computing-intensive and time-demanding task to perform. It is also difficult to prove which interactions will be of particular importance for the investigated property in question, e.g., in vitro or in vivo activity of some sort.

An alternative, more pragmatic, use of the knowledge about the active site is to use it as an "alignment protocol." Here one would start out by docking the ligand and the protein using one of several available docking schemes in a more or less rigid manner followed by the attempt to develop a more quantitative 3D structure–activity model (Goodford 1985; Cramer et al. 1988; Wade and Goodford 1993) based on the resulting orientations of the ligands with respect to the protein. However, this is not a straightforward process since it is known that small variations in the alignments used for each compound in the analysis may greatly influence the quality and predictivity of the derived quantitative structure–activity relationships (QSAR) model (Folkers et al. 1993; Klebe 1993). Still, the approach, together with other attempts conducted at the same time (see below), is a promising one in order to derive a useful and predictive tool for lead structure generation. Lead optimization will probably have to be performed with a more classical approach using a carefully designed set of analogues.

In the absence of a 3D structure of the target protein, or as a simultaneous activity to DDD, IDD is used to generate models both for lead generation and for lead optimization. Karo Bio currently uses the Catalyst software (BioCAD 1994) to generate quantitative models for lead generation. These models, hypotheses in Catalyst terminology, can subsequently be used for 3D data base searches in commercially available data bases as well as in corporate data bases. There are several

Table 7. Quantitative structure–activity relationships (QSAR)

---

1. 2D QSAR
   Free-Wilson type
   Hansch type
   Multiple linear regression techniques
   Partial least squares methodology
   Parallel models
2. 3D QSAR
   Catalyst
   Grid based methods/PLS

---

other programs available for generating pharmacophores such as DISCO (Martin et al. 1993), APEX-3D (Golender and Vorpagel 1993), and AUTOFIT (Kato et al. 1992). However, most of these approaches do not use the investigated property, e.g., biological activity, in an explicit manner during model generation, which may be of considerable importance in deriving correct alignments of a set of structurally diverse compounds. There is also the question of whether to use a certain reference structure or a set of such structures during model generation. It is probably an advantage to use the latter approach since there is usually not one "perfect" molecule to perform the matching against. This is of particular interest in the case of lead generation since here one is dealing with compounds of more or less structural diversity, not wanting to end up in a "me too" situation.

However, for the next phase, that of lead optimization, more traditional QSAR approaches become of interest so as to be able to investigate the finer details, e.g., substitution pattern and lengths of side chains etc., of a promising lead structure. Such techniques may incorporate 3D QSARs such as CoMFA (Cramer et al. 1988) and GRID (Goodford 1985; Wade and Goodford 1993) related methodologies as well as Free-Wilson (FW) and Hansch-like approaches (Kubinyi 1993).

At Karo Bio we use both 3D QSARs of the CoMFA type with software which has been developed in house (Norinder 1993) and 2D QSARs of the FW and Hansch type and combinations thereof. We feel that it is important to work on a particular problem with several methodologies at the same time (Table 7). The different models, which may be more or less obvious to interpret in terms of the descriptors included, i.e., appear as white, gray, or black boxes, may also, at best, interact in a synergistic manner. One model based, for instance, on descriptors of the FW type, may indicate the importance of a particular kind of variable that needs to be included in a successful model. Thus, another model, perhaps based on calculated descriptors of quantum mechanical (QM) type, also have to include such corresponding variables within the QM framework to be good and predictive.

Furthermore, one never knows which model will prove to be the best in terms of predictivity. This may also change as the various models develop over time. Additionally, from the different models a more complete picture of what is important for the target in question may emerge after the initial phases/cycles of model development.

However, as pointed out by others (Pleiss and Unger 1990; Plummer 1990), every good model needs to be based on a training set which covers the available chemical and structural space in a balanced manner. At Karo Bio we use experimental design techniques of the factorial type (Box et al. 1978) to achieve this objective. This strategy is used not only in drug design research at Karo Bio, but also in most other research activities that the company is engaged in where there is a need to obtain an effective, and preferably robust, procedure with a relatively small number of experiments.

Principal component-based methods such as partial least squares (PLS; Wold et al. 1984) are used in QSAR research and most other statistical modelling initiatives as well at Karo Bio. We think that these methods have advantages over traditional multiple linear regression techniques (Cramer 1993) and that these advantages should not be overlooked when trying to develop a good statistical model with reasonable predictivity.

6.5 Limitations

Although rational drug design seems as a quite promising and time- and cost-saving approach there certainly are limitations to what one can expect from CADD. One of these limitations lies in the attempt to rationalize a complex biological (test) system using, in this context, rather simple equations and extensive approximations. In most modeling exercises factors such as dynamics, the influence of a solvent, and entropic effects are only accounted for in a very limited way. In QSAR applications a static situation prevails in most cases and a relatively simple description of the investigated system is usually used. This, in turn, means that the derived statistical relationships must be viewed more as local mathematical constructs with limited applicability (local models). Nevertheless, the use of rational drug design in pharmaceutical research has proven its value and produced some "success stories" in therapeutic areas such as hypertension, cancer, and AIDS.

References

BioCAD (1994) 555 Oakmead Parkway, Sunnyvale, CA 94086–4023, USA

Box GED, Hunter WG, Hunter JS (1978) Statistics for experimenters. Wiley, New York

Cramer RD III (1993) Partial Least Squares (PLS): its strengths and limitations. Perspect Drug Discov Design 1:269–278

Cramer RD III, Patterson DE, Bunce JD (1988) Comparative Molecular Field Analysis (CoMFA). 1. Effect of shape on binding of steroids to carrier proteins. J Am Chem Soc 110:5959–5967

Folkers G, Merz A, Rognan (1993) CoMFA: scope and limitations. In: Kubinyi H (Ed) 3D QSAR in drug design. Escom, Leiden, pp 583–618

Golender VE, Vorpagel ER (1993) Computer-assisted pharmacophore identification. In: Kubinyi H (ed) 3D QSAR in drug design. Escom, Leiden, pp 137–149

Goodford PJ (1985) A computational procedure for determining energetically favorable binding sites on biologically important macromolecules. J Med Chem 28:849–857

Greer J, Erickson JW, Baldwin JJ, Varney MD (1994) Application of the three-dimensional structures of protein target molecules in structure-based drug design. J Med Chem 37:1035–1054

Kato Y, Inoue A, Yamada M, Tomioka N, Itai A (1992) Automatic superposition of drug molecules based on their common receptor site. J Comput Aided Mol Design 6:475–486

Klebe G (1993) Structural alignment of molecules. In: Kubinyi H (ed) 3D QSAR in drug design. Escom, Leiden, pp 173–199

Kubinyi H (1993) QSAR: Hansch analysis and related approaches. In: Mannhold R, Krogsgaard-Larsen P, Timmerman H (eds) Methods and principles in medicinal chemistry, vol 1. VCH Publishers, Weinheim

Martin YC, Bures MG, Danaher EA, DeLazzer J, Lico I, Pavlik PA (1993) A fast new approach to pharmacophore mapping and its application to dopaminergic and benzodiazepine agonists. J Comput Aided Mol Design 7:83–102

Navia MA, Peattie DA (1993) Structure-based drug design: applications in immunopharmacology and immunosuppresion. TIPS 14:189–195

Norinder U (1993) A PLS QSAR analysis using 3D generated aromatic descriptors of principal property type: application to some dopamine D2 benzamide antagonists. J Comput Aided Mol Design 7:671–682

Pleiss MA, Unger SH (1990) The design of test series and the significance of QSAR relationships. In: Sammes PG, Taylor JD (eds) Comprehensive medicinal chemistry, vol 4. Pergamon, Oxford, pp 561–587

Plummer EL (1990) The application of quantitative design strategies in pesticide discovery. In: Lipkowitz KB, Boyd DB (eds) Reviews in computational chemistry. VCH Publishers, Weinheim, pp 119–168

Reich SH, Webber SE (1993) Structure-based drug design (SBDD): every picture tells a story... Perspect Drug Discov Design 1:371–390

Wade RC, Goodford PJ (1993) Further development of hydrogen bond functions for use in determining energetically favorable sites on molecules of known structure. 2. Ligand probe groups with the ability to form more than one hydrogen bond. J Med Chem 36:148–156

Wold S, Albano C, Dunn W III, Edlund U, Esbensen K, Geladi P, Hellberg S, Johannson E, Lindberg W, Sjöström M (1984) Multivariate data analysis in chemistry. In: Kowalski BR (ed) Chemometrics-mathematics and statistics in chemistry. Reidel, Dordrecht, pp 17-95

7 Screening Three-Dimensional Databases for Lead Finding

H. P. Weber

7.1 Introduction

Most drugs currently in use were developed based on optimization of *lead compounds.* The source of lead compounds used to be primarily natural products from plants, microorganisms, and higher species, including hormones, transmitters etc., which were discovered by screening techniques. This approach to drug development was highly successful and is still being successfully applied in most pharmaceutical companies. However, the rapid development of biotechnology has offered other promising routes towards the same goal; the present chapter discusses one of these new routes.

Biotechnology can provide – *in principle* – material for any protein target whose gene has been identified, and modern X-ray crystallography can, *again, in principle,* determine the crystalline structure at near atomic resolution for any protein of therapeutic interest. This allows a direct approach to the design, or search, of biologically active compounds, i.e., the structure-based drug design.

With the knowledge of the target protein in atomic detail, there are two basically different strategies which can be applied in structure-based drug design:

1. Design of a novel ligand for a receptor binding site by applying (graphical) molecular modeling methods. This approach, which involves mainly the interactive use of modeling systems, may be supported by special programs to place small to medium fragments into strategic positions in the binding site, e.g., making favorable hydrogen bonds, and extending the fragment with the same strategy to a "receptor site-filling" molecule. Among the many specialized computing tools for this purpose a few are particularly helpful, e.g., programs such as LUDI (Böhm 1992), CLIX (Lawrence and Davis 1992), or CAVEAT (Lauri and Bartlett 1994). The special character of this approach is the direct control and navigation of all steps in the design of one or a few active molecules by the medicinal chemist. There is some automation involved in the process but is mostly based on individual know-how and expertise in graphical structure-based drug design.

2. The second route to structure-based drug discovery is electronic screening of three-dimensional (3D) databases. In this approach, a large 3D database containing organic compounds is screened by a computer program which tests each compound in the database in an automated fashion as to how well it would fit into a defined receptor binding site and to calculate a score according to the fit. The top-scoring compounds are potential candidates for biological testing.

In this context focus will be on lead finding by electronic screening of 3D databases, although the other approach, molecular modeling, would be equally important and interesting. In fact, the two methods are complementary and have a lot in common.

The main topics to be discussed are:

- 3D databases
  1. Generation of 3D databases from 2D databases
  2. Conformational flexibility in 3D databases
- Defining the receptor binding site
- Docking procedures
- Scoring of docked ligands
- Postprocessing of electronic screening results
- Some computing aspects
- Future aspects and potential in pharmaceutical research

The pioneering work in this field was done by I.D. Kuntz et al. (1982) with the development of a computer program called DOCK. Kuntz was well ahead of his time, but a few years later derivatives of his work quickly followed, e.g., ALADDIN (Van Drie et al. 1989) and FLOG (Miller et al. 1994). The approach has now attracted a good deal of interest by pharmaceutical companies since two of the prerequisites for electronic database screening are available now: (1) large chemical databases, in electronic format, containing all (or a large part of) the compounds which have been synthesized in the company since its foundation, and (2) the biotechnology and the crystallographic know-how to establish 3D models of any protein targets of interest, either in-house or in collaboration with an external company.

One of the prerequisites for this approach is knowledge of the 3D structure of the receptor of interest in atomic detail. Although the biotechnological production of a protein and its structural elucidation by X-ray analysis or by nuclear magnetic resonance (NMR) are major research projects, this aspect will not be discussed here any further. Just one comment shall be made concerning an obvious detail that is sometimes overlooked: the timing. When a new, therapeutically interesting target protein has been identified, a time period of about 2 years has to be anticipated to produce the protein using biotechnological methods and to do the structure using crystallography (or NMR). Only then can structure-based design start. This period may be shortened if, for example, the protein is already physically available, or if a reliable model can be built by homology from a related, known protein structure. Otherwise, 2 years is realistic for planning a novel structure-based project.

7.2 3D Databases

There are a number of "experimental" 3D databases publicly available, which need not be generated. The most prominent among these are the Cambridge Crystal Database (CSD), containing the collected results of some 120 000 published X-ray crystal structure analyses, and the Brookhaven Protein Data Base (PDB) containing about 2500 entries of protein crystal structures (and some NMR solution structures). Both 3D databases are most valuable sources for reliable structural data on both organic and biopolymer molecular conformations as they exist in the solid state.

For the purpose of electronic screening the CSD can be a useful 3D database for lead finding; it contains a broad spectrum of diverse molecular structures in reliable low-energy conformations. However, its use as a source for lead compounds for pharmaceutical purposes is somewhat limited for three reasons: (1) many of the CSD compounds may have an "unbiological chemistry," e.g., undesirable physical or chemical properties such as containing metals; (2) many other compounds of molecules which are interesting a priori are chemically "corrupted," i.e., derivatives made for purely crystallographic purposes; and, finally, (3) most of the compounds will either be unavailable or difficult to get.

The PDB database is, of course, the primary source for the receptor structures. However, a derivative of the PDB database might also be a useful source of potential lead candidates: peptides in the conformation as they exist in proteins, particularly in surface loops, may be interesting peptide leads, since they are in a conformation which, biologically, may be more relevant than the conformation of the same peptides as, for example, found in a crystal structure or modeled in vacuo. The preparation of a 3D database containing such peptides could be a very valuable ligand source in structure-based drug design. Besides these "experimental" 3D databases, there are a few "theoretical," or "modelled" 3D databases. Four such 3D databases, commercially (publicly) available, do exist and are useful for lead finding. Molecular Design Ltd. has prepared and maintains three such modeled databases: the MACCS Drug Data Report file (MDDR, containing some 47 000 compounds), the Available Chemicals Database (ACD, containing about 130 000 commercially available compounds), and the Compounds of Medicinal Chemistry (CMC, with approx. 6500 compounds). *Chemical Abstracts* has also created a huge 3D database of about 6 million compounds in its primary database. The methods and problems involved in creating such "modeled" 3D databases are briefly discussed.

7.2.1 Generation of 3D Databases from 2D Databases

Most pharmaceutical companies, and some university institutes, have large proprietary 2D chemical databases. These databases, originally in a hard copy registry form and now converted into electronic form containing a full description of each compound by chemical topology (CT), bond type (BT) and chirality information, may well contain over 100 000 compounds, and they represent one of the company's (or institute's) most valuable treasures: the compiled and electronically accessible results of the work of many hundreds of medicinal chemists, biologists, and pharmacologists, accumulated in many years of research. Obviously, it is of utmost importance to pharmaceutical research to reevaluate these data as new assays and drug design projects emerge because (1) these compounds are (usually) available in house, and (2) because most of them are compounds of biological potential.

For the purpose of electronic screening, these 2D databases have to be converted into 3D databases by automated procedures. The basic requirements for such procedures are to be robust (i.e., they should not crash with an unexpected type of molecule), to produce stereochemically acceptable and thermodynamically favorable conformations, and to be fast, i.e., using less than, say, 1 cpu/s per 2D to 3D conversion. Two such types of programs have been developed:

1. Programs applying basic stereochemical principles to build a rough 3D molecular model of the compound atom by atom using the stored CT, BT, and chirality information, followed by an abbreviated form of geometry optimization to straighten out the somewhat rough conformation. Popular programs of this type are, for example, CONCORD (Pearlman 1987, 1993) and CORINA (Gasteiger et al. 1990), which both meet the criteria mentioned above. Problems only arise with larger cyclic compounds (higher than 7-membered rings) and with occasional overlap of topologically distant atoms.

2. Programs using a database of fragments [e.g., ring systems, (semi)rigid groups etc.] that are able to assemble a full molecule fragment by fragment, applying empirical rules for fragment connection. Typical programs of this type are WIZARD/COBRA (Dolata and Carter 1987), AIMB (Wipke and Hahn 1988), and CHEM-X BUILDER (Davies and Upton 1990). Many of the fragment-joining rules have been derived from crystal structures, and some learning mechanisms are implemented into a knowledge base to produce crystall-like low energy conformations. This approach is potentially superior to the atom-by-atom approach. However, such programs are relatively slow, using typically about 10 times the cpu-time of the atom-by-atom programs, and are not yet as highly developed as to deal with all (or most) classes of organochemical compounds.

Both types of 2D to 3D converting programs have the same limitation: they are able to produce one, and only one, conformation for a particular molecule. This particular conformation may be a stereochemically reasonable one – it may even be a low energy conformation; but the ultimate request for the purpose of electronic screening is to have the conformation of the molecule when bound to the receptor of interest in the database! And for a potentially flexible molecule the "modeled" conformation in the database may well be different from the required one. This basic problem of electronic screening leads to the next point: how to address *conformational flexibility* in electronic database screening.

7.2.2 Conformational Flexibility in 3D Databases

All docking procedures presently in use in electronic screening of 3D databases use the *rigid ligand–rigid receptor docking* approach. An obvious, and at first glance attractive solution to conformational flexibility in this case is to use a 2D to 3D conversion program which will produce multiple conformations of a compound. Clearly, the size of the 3D database would then increase by a factor of, say 20, if on average 20 conformers of a compound were to be generated. With today's computer resources, such a multi-conformer 3D database containing some

1 000 000 original compounds would require about 6 GB of disk space, which is large but realizable. The time to search such a multi-conformer 3D database would, of course, also increase by the same factor. However, as attractive and elegant this way appears to be, the problem of conformational flexibility is only partly solved with such a multiconformer 3D database: clearly, the probability of having the "right" receptor-bound conformation of the molecule among the conformers is now increased, but there is still no guarantee that the "right" one is among them!

To my knowledge, there is only one public program to date, CATALYST (BioCAD Corporation 1994), which produces up to a user-defined number of conformers per compound, whose internal energy lies within a defined energy range. In our laboratory we have converted a 2D database of some 150 000 compounds with CATALYST (Version 2.0) into a multiconformation 3D database. The quality of the conformers generated is difficult to assess objectively, but from looking at some of the structures in our 3D databases at random one gets the impression that in general the conformations which have been generated are reasonable and conformational space is well explored, but some badly distorted ring structures which were found have cast some doubt on the quality of their algorithms and we suspect that there are still some systematic faults in either recognition or building of molecular conformers. However, since the authors of CATALYST do not provide detailed documentation on their methods, either for how they generate the starting conformation or how they explore conformational space for diverse conformers, it is not possible to say much more at this point.

Another solution to conformational flexibility would be to generate diverse conformers "on the fly" during the docking procedure. There is an interesting approach already in use for flexible ligand searching by TRIPOS (1994): starting with a random (but reasonable) conformation of a molecule (as it is in the 3D database), the algorithm tries to change the conformation in a directed way through bond rotations (excluding ring bonds!) so as to achieve a conformation satisfying some *predefined constraints* without producing excessive internal strain. Clearly, the ideal application of this method is the search for compounds containing a particular *pharmacophoric* pattern, defined with interatomic distances and angles. However, in the case of 3D database screening for compounds fitting into a receptor binding site, it is usually difficult to define distance or angle constraints in advance, and the method may only find application in special cases of electronic screening.

In summary, the approaches as discussed above do not provide a satisfactory solution to the problem of conformational flexibility. At present, the (rigid) multiple conformer 3D database seems to be the best and most practical way available. New original approaches have still to be developed.

7.3 Defining the Receptor Binding Site

Let us assume that a 3D database and a binding site in a protein (in atomic detail) are at hand. The next problem, then, is to characterize the binding cavity in such a way that computational docking of a compound (from the database) into the binding site can be done. The basic idea of how to do this is to place a number of discrete, strategically distributed points into the receptor cavity, forming the so-called "cluster of centers," and then try to fit the atoms of a ligand onto these points.

In the pioneering work of Kuntz et al. (1982), a general method of how to produce such a cluster of centers was proposed, based on the use of Connolly's (1983) solvent-accessible surface. It proved to be a very useful method and he and his collaborators were able to show that with their method some crystallographically determined protein–ligand complexes could be reproduced quite accurately (DesJarlais et al. 1988).

Many refinements and variations of his original method have subsequently been proposed. One of these variants developed in our laboratory (P. Burkhard, unpublished work) places the centers in closest contact with the receptor surface, postulating that fitting ligand atoms to such points will also produce the most effective fit of a ligand. These closest points are situated at approximately double the *van der Waals* distance from receptor surface atoms, i.e., they are initially close to or on the *Lee-Richards surface* of the receptor binding site (Lee and Richards 1971). In fact we used a *modified Lee-Richards* surface, which is obtained with a probe atom whose radius varies with the receptor atom in contact (assuming that the best contacts between ligand and receptor atoms occur when both are of similar type) as, for example, given in Table 1.

Table 1. Ligand and receptor atom radii

Receptor atom	Receptor atom radius	Ligand atom	Ligand atom radius	Double vdW radius
C	1.7	C	1.7	3.74
0	1.4	N/O	1.5/1.4	2.85
N	1.5	0	1.4	2.90
S	1.9	Any	1.4–1.7	3.60

Since hydrogen atoms are ignored, the C...C vdW radius was somewhat increased.

Other schemes of defining a cluster of centers for the receptor site are using points of an evenly spaced potential energy grid put over the site; on every grid point the interaction energies of various probe atoms with the receptor are calculated. The grid points with the highest interaction energy may then be retained as centers defining the receptor site (Miller et al. 1994).

Clearly, a good definition of the receptor binding site is an important aspect of such docking procedures and great care should be taken in this step; the choice of method will also depend to some extent on particular features of the receptor binding site of interest, e.g., if it is a deep hydrophobic pocket or a rather shallow cavity with polar residues. Some of these considerations are discussed in the next section.

7.4 Docking Procedures

The fundamental problem of docking a ligand molecule into a receptor cavity is to fit n ligand atoms (out of N_l atoms in the ligand) onto n centers in the receptor binding site (out of N_r receptor centers). This is a formidable combinatorial problem. The number of possible pairing sets Z is

$$Z = {}^nC(N_r) \times {}^nP(N_l) \sim (N_r)^n \times (N_l)^n$$

C and P represent combinations and permutations, respectively (Shoichet et al. 1991). The approximation is valid in case of $n \ll N_r$, N_l, which in practice is normally the case, since the number of ligand

Table 2. Heuristics for rigid ligand–rigid receptor docking procedures

(0)	$k = 0$ *(k, l are receptor center numbers)*	
(1)	$k = > k + 1$	Loop on centers
	Select (and flag as used) receptor center k	
	If $k > Nr$, then continue procedure with the next ligand (0)	
	$i = 0$ *(i, j are ligand atom numbers)*	
(2)	$n = 0$ (number of matching intraligand/intracenter vectors pairs found)	
	$i = > i + 1$	Loop on atoms
	Select (and flag as used) ligand atom i	
(3)	Select the longest intraligand distance d_{ij}	j not flagged
	Search the best-matching intracenter distance d_{kl}	l not flagged
	If $\mid d_{kl} - d_{ij} \mid < e_d$, then $n = > n + 1$, else continue with phase (2)	
	Check on the *(n–1)* "cross vector differences"	
	If all *(n–1)* cross-vector-differences $< e_c$	
	then flag ligand atom *j and receptor center l, j = >* i,	
	continue with phase (4), else continue with phase (1)	
(4)	If $n = n_{max}$, then phase (5), else continue with phase (3)	
(5)	Found one match:	
	Do docking	
	Do scoring, store, and restart with phase (2)	

atom–receptor center pairs, n, to be matched, usually 4 (minimum) to 8, is much smaller than the number of atoms in the ligand (typically 30) or centers in the binding site (typically 100–200). In any case, the number of Z is so enormous that a systematic and exhaustive docking procedure is beyond any practical relevance.

Therefore some heuristic approaches to this combinatorial problem have to be applied: for detailed descriptions see, for example, Shoichet et al. (1991), or Miller et al. (1994). In Table 2 we summarize a general scheme of the heuristics as implemented in most docking procedures.

Variations in this procedure can be introduced at all stages. An interesting modification for example can be introduced at stage 1 as follows (Lee and Richards 1971): selected receptor centers can be given properties, for example, hydrogen bond donor/acceptor property, or lipophilic/hydrophilic property. If receptor centers of a particular property match with ligand atoms of the same property, it would be positively scored, but negatively if they do not match. Or another interesting variation: a few receptor centers (or small clusters of centers) may be assigned to be always occupied with a ligand atom of a particular type, for example, by a hydrogen bond acceptor atom of the ligand so that a hydrogen bond to the receptor is possible. Such modifications will speed up the procedure tremendously by rejecting at an early check many of the docking attempts. Such restrictions will produce far fewer successful dockings than without, but all of the successful dockings would each show the particular required feature.

In stage 3, also many variations are possible; for example, the selection of the next intraligand distance need not be the next longest, but could be chosen so as to be at maximum distance from all previous points. Or, variation in the tolerance for distance checks (ed *and* e, in Table 2) could be introduced, depending on the type of ligand atom and property of the receptor center (Lee and Richards 1971).

A comment to stage 5: after a matching set of intraligand and intracenter distances has successfully been found, a superposition – usually a least squares fit – of the selected ligand atoms onto the matching receptor centers is done. Since up to this point only intraligand and intracenter *distances* have been matched, it is possible that the fit is much worse than one would expect from the tolerances *(ed, ec)*, which would indicate that the enantiomer of the original ligand should be fitted.

7.5 Scoring of a Docked Ligand

After a successful docking has been achieved, the score must be calculated to quantitatively assess the "goodness of fit." A good scoring function should take into account (1) *the shape fitting* of ligand and receptor, which basically is a measure of the contact surface, and (2) the "chemical fit," i.e., check of the presence of matching or nonmatching ligand atom/receptor atom properties, respectively, such as whether contacting hydrogen bond partners are of the donor/acceptor type, or whether polar interactions are of opposite charge, or whether lipophilic ligand atoms are in contact with lipophilic receptor atoms, etc.; all these features would be positively scored if true, but negatively if false.

Obviously, the calculation of the *van der Waals* and *coulombic* interaction energy between ligand and receptor would be a good score, assessing both shape and chemical fit quantitatively. The prerequisite for such a scoring calculation is the choice of an appropriate force field and the complete assignment of atom types and of partial atomic charges for ligand and receptor atoms.

An "all-H included" force field, or even an "only polar-H included" force field for scoring would be an adequate choice. However, there is a serious problem with such force fields in current rigid ligand–rigid receptor docking procedures: the position of the free-rotatable OH hydrogens! Assumptions on these positions have to be made prior to docking. This, however, may give a bad score since a OH hydrogen may be pointing towards a hydrogen bond donor and hence give a negative score, although only a small displacement on the hydrogen bond-hydrogen would be needed to restore a good score. This may be so bad that an otherwise good docking position may be lost. The elegant way to eliminate such a situation is, of course, to make a search for the best positioning for every free-rotatable OH hydrogen, or, even better, to do a full energy minimization of the receptor–ligand complex. These additional calculations, however, which would have to be done for every ligand and every fit in turn are computationally so expensive that such a procedure becomes impractical.

A remedy to this situation could be the choice of a "no-H included" force field. Such a force field is certainly a poorer choice than any of the other two force fields, but it effectively eliminates the problem mentioned above. An additional advantage of a "no-H included" force field is that assignments of partial atomic charges to any atoms is not needed, since the calculation of a *coulombic* energy for hydrogen bonds without a hydrogen would be meaningless. We have been working with such a no-H force field and have good experience with the following scoring function:

1. A (negative) *van der Waals* potential is calculated between all (non-hydrogen) atoms of the ligand and of the receptor binding site (with a receptor atoms cutoff distance of, for example, 6 Å)
2. A *hydrophobicity* and *hydrogen-bond* scoring function of the form

$$C \cdot \exp[-A(d - d_{opt})^2]$$

where C, A, and d_{opt} are parameters depending on the atom types, and d is the distance between a pair of ligand and receptor atoms.

A very interesting scoring function has recently been published by Böhm (1992), where the author proposes a function to estimate the ligand–receptor affinity. A relatively simple approach to estimate $\Delta G_{binding}$ was proposed, consisting of only five terms:

1. A basic ground value, ΔG_O, independent of any interactions
2. A term ΔG_{HB}, describing the contribution of hydrogen bonds (depending on the geometry of the hydrogen bond: $d = d(D \dots A)$, $\alpha = (D - H..A)$
3. A term for lipophilic interactions, ΔG_{lipo}, depending on the lipophilic contact surface
4. A term for ionic interactions, ΔG_{ionic}, depending on the geometry of the interaction
5. A term accounting for freezing internal degrees of freedom in the ligand, ΔG_{rot}, linearly depending on the number of rotational bonds

The few parameters in the equation:

$$\Delta G_{binding} = \Delta G_o + \Delta G_{HB}\, f(d,\alpha) + \Delta G_{lipo}\, A + \Delta G_{ionic}\, f(d,a) + \Delta G_{rot}\, N_{rot}$$

were calibrated using some 40 ligand–protein complexes from the Brookhaven database with published K_D values. The cross-validated r^2 for the resulting function was 0.696, i.e., the predictive power for $\Delta G_{binding}$ was about 2.3 kcal/mol, corresponding to about 1.6 in K_D. The simplicity of the estimate and the speed with which it can be calculated at the end of a ligand docking make this scoring function look very promising. Further refinement of the method has been announced.

Another interesting development which recently has been published (Miller et al. 1994) concerns optimization of the fitted ligand after successful initial docking: as pointed out above, a straight energy minimization of the ligand–receptor complex is impractical in screening large databases due to exessive computing time. However, Miller et al. (1994) proposed a simplex optimization of the rigid ligand in the (rigid) receptor, which significantly improves the scoring with only modest additional cpu time .

Most newer DOCK programs are using a potential energy grid for the calculation of the ligand–receptor interaction energy. The basic idea is to precalculate the interaction energy of the receptor with various atom types placed in turn on the grid points of a grid covering the binding site, i.e., a 3D table containing an interaction energy vector on each grid point. The score for a fitted ligand is then a simple sum of the appropriate values from the interaction energy vectors at the grid points closest to the ligand atoms. A 3D interpolation between the ligand atom and the eight nearest grid points can be used for higher precision, which will, of course, also be achieved by reducing the grid spacing. The size of the grid and available memory size will ultimately define the choice of the grid spacing (typically .2–.5 Å).

7.6 Postprocessing of the Electronic Screening Results

The primary postprocessing of the docking results is, of course, a sorting according to the scores. If the scoring function were absolutely reliable, and if the docking procedures were perfect, too, the top-scoring compounds could go straight for biological testing without any further inspection. But in actual practice the, say 300 top-scoring compounds have to be inspected graphically/visually by a medicinal chemist for various aspects such as how the conformation of the ligand is, how the ligand is docked into the binding site, what kind of interactions exist between ligand and receptor, and to what chemical class the ligand belongs, etc. This is both a tedious and subjective process. Clearly, automated postprocessing would be useful.

Automated postprocessing could, for example, be applied to collect families of chemically similar ligands, so that one could visualize the best representatives family by family and select the most promising candidates for testing from each family. This would also improve the chances of finding chemically diverse new leads, rather than wasting time assaying dozens of compounds which are all basically alike.

Another approach, which might be called preprocessing, would be to re-sort the original, large 3D database according to chemical diversity and work with a subset of, say, the 10% most diverse compounds. This would have the additional advantage of reducing the computing time (which might be used for improved scoring and optimizing ligand fitting).

However, there may always have to be a final visual (graphical) inspection of the top-scoring candidates, since some of the criteria for selection must be assessed by the medicinal chemist himself, e.g. solubility of a compound, chemical stability, synthetic possibilities of derivation, chemical intuition – in short, criteria which are difficult to program!

7.7 Some Computing Aspects

The part which is computationally most demanding in electronic screening of large 3D databases with one of the methods described is the docking procedure. Improvements in efficiency without sacrificing too much accuracy is the goal. The main parameters governing this part are: (a) the number of centers defining the receptor binding site – clearly, the higher the density of these centers, the better the chances of finding matching distances, but also the higher the cpu time consumed per ligand; (b) the heuristics of matching intraligand distances to intracenter distances – in particular, specifying receptor centers which have to be occupied by ligand atoms at the start of the docking procedure can greatly reduce the computing time. The second demanding aspect of computing is the scoring. In this phase, the "potential energy grid" concept is much more efficient than a repetitive *van der Waals* or *coulombic* interaction energy calculation. Implementing a short (rigid) ligand position optimization before scoring, or even a partial energy relaxation of ligand and receptor, will add some computing time but improve on scoring (Miller et al. 1994; Meng et al. 1993).

Although optimizing a DOCK program according to the points above will improve performance, the most effective, and fortunately easiest way to speed up will be achievd by parallelization of the whole process. The nature of the procedure is such that this can be done in either of two ways: operating on one single database, and forking into child processes, each handling one ligand at the time with a master process watching over the number of active child processes and opening a new one as soon as one dies, or splitting the original database into say 12 subsets and have 12 processors running simultaneously, each working on a different subset. Both methods have advantages and which one is chosen depends on the type of computing facility available.

7.8 Future Aspects and Potential in Pharmaceutical Research

The most severe deficiency of the present DOCK methods is their restriction to "rigid ligand to rigid receptor" fitting. While a multiconformer database relaxes this limitation at least partially on one side, no practical method has yet emerged to deal with a fully flexible system.

One single, exploratory experiment was done in our laboratory with a "flexible ligand to (semi)flexible receptor" fit, using the modified X-PLOR scripts *random* and *sa* (Brunger 1992). The first script randomizes the coordinates of the ligand and the second reconstitutes a molecular conformation by simulated annealing procedures. To ensure that the molecule is generated within the binding cavity the center of mass was tethered to some point in the middle of the cavity and the side chains lining the cavity were left flexible. Such a procedure was successful in regenerating correctly the X-ray structure of a HIV-P–inhibitor complex by restarting the script 30 times, which took 30 h of cpu time on an Alliant FX2800 using eight processors in parallel. It was clearly a success, but at an enormous expense of computer time, and hardly a direction to follow in the future for screening 3D databases.

Since no real breakthrough of practical relevance for flexible ligand flexible receptor docking is in sight, the impact of electronic screening in pharmaceutical research may be for the time being somewhat reduced. However, what is available and practical should be used and applied; only experience will tell how useful the methods are, regardless of their scientifically unsatisfactory state. It is up to the experts to keep expectations at a realistic level and to motivate further development.

With today's docking methods, the best results may be expected by use of a multiconformer 3D database, applying state-of-the-art rigid ligand to rigid receptor docking, followed by a fit optimzation, and using a scoring function on the base of a force field energy grid (or an affinity scoring as proposed by Böhm 1992). These procedures will work best if the receptor binding site is a deep and and almost closed cavity, such as is the case for many enzymes (e.g., HIV-protease, thrombin, sialidase, etc.). In most of these these cases, the side chains in the cavities are tightly packed, and the "rigid receptor" restriction is an adequate assumption. The worst case for the present methods is the search of ligands for a rather flat binding site (e.g., above a β-sheet), lined with long and polar side chains which are floppy and can move independently. An example of this situation is the putative binding surface of the coreceptors CD2/CD58. There are, however, also inter-mediate cases with a shallow, but still distinct binding cavity lined with more or less tightly packed side chains, e.g., cyclophilin, or FKBP12.

One last comment: Electronic screening of 3D databases has to compete with biological screening. Both electronic and biological screening have the same purpose: to discover new leads. Electronic screening relies on 3D databases, powerful computing resources, and a target structure known in atomic detail. Application is relatively cheap and fast, but it produces "theoretical" results of limited reliability, which first have to be tested experimentally before being useful. Biological screening, in contrast, relies on the availability of a "physical" compound store (i.e., a vast collection of diverse compounds to test), and of highly automated assays. Its application is releatively expensive and slow, but it produces experimental results which are of immediate use. Rather than being competitors, I think that the two drug discovery approaches are complementary; while electronic screening provides potentially active candidates for special targets to be tested in special assays, the biological screening is searching for leads of a more general interest with a battery of general assays of high capacity and automation.

References

BioCAD Corporation (1994) Sunnyvale, CA, CATALYST documentation version 2.1

Böhm HJ (1992) The computer program LUDI: a new method for the de novo design of enzyme inhibitors. J Comput Aided Mol Design 6:593–601

Brunger A (1992) X-PLOR version 3.1. Yale University Press, New Haven

Connolly ML (1983) Analytical molecular surface calculation. J Appl Cryst 16:548–558

Davies K, Upton R (1990) Experiences building and searching the Chapman and Hall dictionary of drugs. Tetrahedron Comput Methodol 3:665–681

DesJarlais RL, Sheridan RP, Seibel GL, Dixon JS, Kuntz D (1988) DOCK: a program to search molecule databases. J Med Chem 31:722–734

Dolata DP, Carter RE (1987) WIZARD: applications of expert system techniques to conformational analysis. 1. The basic algorithms exemplified on simple hydrocarbons. J Chem Info Comput Sci 27:36–45

Gasteiger J, Rudolph C, Sadowski J (1990) Automatic generation of 3D-atomic coordinates for organic molecules. Tetrahedron Comput Method 3:537

Kuntz D, Baney JM, Oatley SJ, Langridge R, Ferrin TE (1982) A geometric approach to macromolecule – ligand interactions. J Mol Biol 161:269–288

Lauri G, Bartlett PA (1994) CAVEAT: a program to facilitate the design of organic molecules. J Comput Aided Mol Design 8:51–66

Lawrence MC, Davis PC (1992) CLIX: a search algorithm for finding novel ligands capable of binding proteins of known three-dimensional structure. Proteins Struct Funct Genet 12:31–42

Lee B, Richards FM (1971) The interpretation of protein structures: estimation of static accessibility. J Mol Biol 55:379–400

Meng CM, Gschwend A, Blaney JM, Kuntz ID (1993) Orientational sampling and rigid body minimization in molecular docking. Proteins 17:266–278

Miller MD, Kearsley DJ, Underwood DJ, Sheridan RP (1994) FLOG: a system to select 'quasi-flexible' ligands complementary to a receptor of known three-dimensional structure. J Comput Aided Mol Design 8:153–174

Pearlman RS (1987) CONCORD: from connectivity to 3D-coordinates of organic compounds. Chem Design Auto News 2:1–14

Pearlman RS (1993) Three-dimensional structures: how do we generate them and what can we do with them? Chem Design Auto News 8:3–15

Shoichet BK, Bodian DL, Kuntz ID (1991) New methods for matching ligands to the receptor sites. J Comput Chem 13:380–397

TRIPOS (1994), St Louis, MO UNITY 2D13D, a 3D-database management system

Van Drie JH, Weininger H, Martin YC (1989) ALADDIN: an integrated tool for computer-assisted molecular design and pharmacophore recognition from geometric, steric, and substructure searching of three-dimensional molecular structures. J Comput Aided Mol Design 3:225–237

Wipke WT, Hahn MA (1988) Artificial intelligence model builder. Tetrahedron Comput Methodol 1:141–153

8 Optimization of Peptide Leads and Molecular Modeling

J. L Fauchère

8.1 Introduction

Since the early 1960s when chemists demonstrated their ability to reproduce by chemical synthesis peptides of up to 40 amino acids, a great deal of experience has accumulated in terms of modification and optimization of natural peptides. The purpose of optimization has been both to design selective pharmacological tools as receptor probes and to obtain useful therapeutic agents. Impressive increases in potency, selectivity, and stability of peptide leads have been achieved, including the design of peptide-derived nonpeptide mimetics. In the past few years, new sources of peptide leads have appeared and the technologies of peptide drug design have gained considerably in efficacy. The following short overview is an attempt to analyze the success and potential of these methods and to critically evaluate the specific contribution of molecular modeling and computational methods to the overall optimization process.

8.2 The Sources of Leads: Natural Peptides and Peptide Libraries

Drug discovery often relies on the traditionally recognized therapeutic virtues of a given natural compound. Peptide drug design also follows this path, starting with natural peptides obtained from the traditional sources of drugs (microorganisms, plant and animal cells), mostly on the basis of a biological assay. An interesting, rich source of peptides is the skin of amphibians (Erspamer and Melchiorri 1980; Barra et al. 1994), since most of these peptides have phylogenetic parent counterparts in the gut and the brain. Chemically speaking, peptides are relatively short chains of amino acids (compared to proteins), in which these chiral building blocks are attached through a planar amide bond, mostly as *trans* substituents. In particular, peptides are found as hormones, neurotransmitters, cytokines, growth factors, or inhibitors of proteases in the human body, in which they act much like a drug, exerting a strong and short effect before being inactivated via proteolysis to nontoxic metabolites, the amino acids. Biologically active peptides are generally contained as cryptic inactive sequences in protein precursors, from which they are liberated by the action of specific processing proteases, thus giving rise to the active species. New peptides and peptidelike compounds are being discovered at a high rate in the brain, other organs, and body fluids, thus providing a rich source of molecular diversity.

Peptide libraries are another relatively new source of peptide leads, since they contain millions of short peptides, typically hexapeptides, which can be screened in a biochemical, pharmacological, or immunological assay (reviews: Gallop et al. 1994; Gordon et al. 1994). These libraries are generated, for example, by chemical synthesis on solid phase, in a combinatorial way by a "mix and divide" procedure after each incorporation of a new amino acid (review: Houghten 1993). Similar libraries can also be obtained at the surface of phages after incorporation of random codons in the genetic material (Scott and Smith 1990). The production of peptide libraries has initiated the broader process of generation of molecular diversity (review: Moos et al. 1993) by repetitive random synthesis, using variable building blocks and variable central scaffolds. Peptide libraries theoretically contain all the possible combinations of the 20 coded amino acid residues (possibly extended to other amino acid building blocks). Thus, in spite of the inherent limitation imposed by the rigidity of the peptide bond (in contrast to the free rotating bonds of psi[CH_2-NH or CH_2-S] pseudo-peptides, for example), such libraries (e.g., of hexapeptides) are likely to provide efficient coverage of the conformational space within the envelope defined by the most extended hexapeptide. A few peptide leads have been discovered by screening of libraries (e.g., O'Neil et al. 1992; Zuckermann et al. 1994).

Several features point out the peptide as an outstanding lead. Its primary structure is generally easy to elucidate by protein sequencing, its synthesis easy to perform by repetitive procedures on solid phase, and its biological activity often known. These are good premises for a chemical optimization of the peptide lead towards a potential thera-peutic agent.

8.3 The Process of Optimization: Chemical Modifications and Structure–Activity Relationships

From the point of view of the medicinal chemist, peptides are ideal leads, but they are poor drugs. Due to their modest bioavailability (resulting from high hydrophilicity) and their short duration of action (resulting from enzymatic degradation in vivo), peptides are limited in their potential for treating a variety of diseases. In contrast, the natural bioactivity of the physiological peptide can easily be followed and its variation used as a guide to establish structure–activity relationships (SAR). There has been a increasing effort to rationally design highly active analogs of biologically significant peptides and the principles which have emerged from these studies have been outlined (Fauchère 1986, 1989, 1994)

Briefly, starting with the shortest sequence which retains appreciable biological activity, chemical modifications are performed in order to improve potency, selectivity, and stability of the original peptide. Single or multiple amino acid substitutions give the first clues. An Ala scan along the sequence often helps to estimate the relative importance of each amino acid side chain (e.g., vasoactive intestinal peptide: O'Donnell et al. 1991). Optimization of the N- and C-terminal ends are attempted to increase both stability and potency. More pronounced modifications tend to restrict the conformational freedom, thus increasing selectivity and suggesting heterocyclic building blocks which could match the peptide cycle. The end compound may be either the original shortened or full-length peptide (thymopentin: Malaise et al. 1985; tetracosactide: Rittel 1973; calcitonin: Azria 1989), a pseudopeptide (goserelin: Dutta et al. 1978; octreotide: Pless et al. 1986; hirudin analog MDL 28050: Krstenansky et al. 1990); renin inhibitor A72517: Rosenberg et al. 1993) or a peptide mimetic (cilazapril: Attwood et al. 1984; RGD mimic: Ku et al. 1993, cyclic urea HIV protease inhibitor DMP323: Lam et al. 1994). Chemically, a combination of classical peptide synthesis on solid phase or in solution and of the traditional methods of organic chemistry is required.

The designer must be assisted by two instruments of the utmost importance during this process. Firstly, estimation of the proteolysis of the original peptide lead and of the subsequent analogs has to be integrated early as a check of stability increase (see Fauchère and Thurieau 1992). Secondly, a simple test of oral bioavailability should be devised to follow the absorption of the most active analogs during optimization (Paladino et al. 1994; Rosenberg et al. 1993).

At this stage, the feeling may be that the tools already exist for a rational optimization of a natural peptide sequence. While from the technical point of view this may be true, the targeted chemical modification of the original lead is still a major challenge, since it is only guided by the resulting values of the biological activity. Obviously, the most difficult part is the selection of the next compound to be synthesized and the prediction of the effects on potency and selectivity.

Classical SAR relies mainly upon the intuition of the chemist to correctly interpret the results. Although this process has proven successful in the past, in regard of the number of achieved peptide optimizations, it is time-consuming and often not far from trial and error.

Quantitative structure–activity relationships (QSAR) studies of peptides are difficult (Pliska 1981; Fauchère 1984; Hellberg et al. 1986; Mager 1994). Provided they are limited to variations of the amino acid side chain in a single position, they may guide design and at least in one case (angiotensin analogs) it could be shown that the same parameters determine both conformation and potency (Fauchère and Lauterwein 1985). However, QSAR generally fails to provide reliable predictions outside the variation domain of the parameters identified in the model. Even comparative molecular field analysis (CoMFA) the most elaborate example available of three-dimensional (3D) QSAR (Cramer et al. 1988; Clark et al. 1990) – in which analogs are aligned with a putative pharmacophore, surrounding fields are calculated and mapped to a 3D-grid, and a QSAR table is generated for partial least squares (PLS: Cramer 1993) analysis – are not always predictive (review: Kubinyi 1993).

For these reasons, great expectations have been put in molecular modeling as a support for a rational optimization of peptide leads.

8.4 Molecular Modeling: Graphics, and Computational Tools

Since no molecular structure conceivably endowed with biological activity can be drawn from scratch on the computer screen, molecular modeling must start from the given structure of either the ligand or the corresponding macromolecular receptor or enzyme. The graphic representation of this structure is the necessary starting point.

Molecular Graphics
The chemist has not awaited the advent of computer display of molecular graphics to rely upon 3D representations of molecules. Ball and stick models (as those constructed by Dreiding) or scale models extended later for proteins to CPK models (Koltun 1965) are examples of widely used stereochemical pictures of molecular structure. However, molecular graphics on the computer screen are far better tools with respect to the ease of construction, dynamic visualization, or superimposition. Graphics can also present the designer with structures that show the van der Waals radii of atoms, the contour of electron density, the solvent-ac-

cessible surface or the Connolly (1983) surface of the molecule. The measurement of interatomic distances and of dihedral angles is greatly simplified. Furthermore, the display of X-ray or nuclear magnetic resonance (NMR)-derived structures of macromolecules cannot be thought of without modern molecular graphics which have become indispensable tools in structure-based design. Display of the secondary structure of peptides (and proteins) are particularly relevant, since the α-helix, the β-pleated sheet or the 3_{10}-helix are common motifs in this class of compounds. Modern graphic 3D representations of the α-helix dipole (Hol et al. 1978), of the amphipatic helix (Edmundson wheel:Schiffer and Edmundson 1967) or of the combination of four such helices as in chymohelizyme (Hahn et al. 1990), or recent graphics of the ion channels found in aibol antibiotics (Toniolo and Benedetti 1991) or in gramicidin A (Wallace and Ravikumar 1988) are impressive and useful descriptions of these biologically functional molecules.

The graphic display of molecular potentials at the surface and in the surrounding space is also commonly used. Molecular potentials share the property of being increasingly "felt" by another approaching molecular species and therefore involved in the molecular interaction. The electrostatic potential (Weiner et al. 1982) is a useful descriptor for the prediction of intermolecular reactivity. It is also largely recognized that the partitioning of a drug between water and octanol often mimics its interaction with the receptor and that biological activity can often be related to lipophilicity (hydrophobicity) as estimated from log P values (Hansch and Leo 1979). It is therefore relevant to model the molecular lipophilicity potential at the molecular surface and in the surrounding space (Fauchère et al. 1988; Furet et al. 1988; Audry et al. 1989, 1992; Kellogg et al. 1991). In addition, Gaillard et al. (1994) have shown that the molecular lipophilicity potential can be used as a third field in CoMFA studies. Its computation and representation are now integrated in most of the software used for peptide studies.

Energy, Molecular Mechanics

Since the 3D structure in vacuo is governed by intramolecular forces which tend to minimize the overall energy of the molecule, a classical mechanical picture has been developed under the general name of molecular mechanics (Buckert and Allinger 1982; Boyd and Lipkowitz 1982; review: Lipkowitz and Boyd 1991). A quantum mechanical treatment of such a complicated system as, for example, a hormone–receptor complex, being out of reach, an empirical function is used which is the sum of the most relevant terms of the intramolecular energy. The latter include electrostatic, van der Waals, dipole–dipole, and bonding potentials. Other potential energy terms may consider hydrogen bonding, solvation, or hydrophobic effects. Despite the proper limitations of the force field employed, molecular mechanics is the basis of most of the software dealing in a simple way with geometry, rotational barriers, steric strain, and conformation. Amber, Charmm, Discover, Gromos, and Sybyl (review: Boyd 1993) are examples of programs which contain force fields most appropriate for the study of peptides, proteins, and other macromolecules.

Molecular mechanics obviously depends on the geometrical construction of the starting molecule. The purpose of avoiding local minima is one of the several reasons why molecular dynamic and Monte Carlo simulations were introduced. Molecular dynamics is the simulation of the dynamic motions of molecules described by the newtonian equations of motion. Obviously, the ability of the system to make transitions over the activation energy barriers depends on temperature. Optimization of the conformation of a macromolecule by simulated annealing (Kirkpatrick et al. 1983) is the classical example of molecular dynamics in which the macromolecule is brought to high temperature and then cooled according to a slow annealing schedule (analogy with annealing in solids), thus increasing the chance of geting the global energy minimum. In the case of Monte Carlo methods, barrier crossing occurs by random generation of configurations and acceptation of higher energy states as a function of temperature. The above software all include a molecular dynamics module well adapted to the study of peptides.

Distance geometry (Crippen 1981) defines molecular geometry by a set of interatomic distances which fully describe conformers. This procedure is useful for receptor site modeling and for the determination of the 3D structure from NMR mesurements (Havel and Wüthrich 1984). Distance geometry approaches the multiple minimum problem differently: since the generated structures may well be energetically unfavorable, the given set of 3D structures is optimized by energy minimization until the lowest energy conformer is reached. The tree search of internal coordinates (Lipton and Still 1988) which generates starting geometries for optimization by molecular mechanics is a convenient alternative algorithm. It is fast since it uses no energy gradient and incorporates a set of criteria to eliminate chemically unreasonable structures in its otherwise exhaustive search for possible conformers. It is particularly useful for automatic handling of the conformational multiplicity of cyclic compounds, a problem solved in only few modeling systems so far.

One extremely useful starting point for molecular modeling is the 3D structure of the lead compound as obtained by NMR spectrosocopy (Kessler et al. 1985; Griesinger et al. 1987). This powerful technique clearly detects hydrogen bonds which can be introduced as constraints in modeling and conserved during optimization. In fact, an average overall 3D structure can often be defined in solution even for complex molecules, which can be a better starting point for modeling. The technique has been applied in conformational studies of CCK_{26-33} (Fournié-Zaluski et al. 1986), of antamanide and somatostatin (Kessler et al. 1986), of a cyclic enkephalin analog (Mammi et al. 1985), of a peptide ligand of pancreatic elastase (Clore et al. 1986), and of a gonadotropin releasing hormone (GnRH) antagonist (Baniak et al. 1987).

Although clear examples of bioactive peptides have been given for which the receptor-bound conformation is significantly different from the conformation in solution (cyclosporin: Weber et al. 1991; Fesik et al. 1991), attempts to detect a pharmacophore are still considered useful. From the point of view of molecular modeling, a pharmacophore can be defined as an arrangement of characteristic atoms in space which is required to trigger the receptor (or to bind to the active site) and which is therefore common to all active species on this receptor. A convincing pharmacophore was detected, for example, in the case of angiotensin-converting enzyme inhibitors, made of three space-oriented determinants: a hydrogen bond acceptor, an anionic salt bridge promoter and a zinc-complexing active site with tightly defined relative distances (review: Hangauer 1989). Three-dimensional superimpositions of energy-minimized ligands acting on the same receptor or enzyme site are of great help in the search for a pharmacophore likely to be operational in drug design.

The number of available crystal structures of enzymes and proteins is increasing rapidly, which opens the way to more and more ligand docking experiments. Apart from manual docking in a convenient graphic system, automatic routines for automatic scanning (Ho and Marshall 1990) are available which can be complemented by intermolecular energy probing (Goodford 1984).

On the whole, the computational tools are now available which render molecular modeling of peptide ligands extremely informative, even in the early steps of optimization where the peptide structure exhibits exceedingly high flexibility. Progressive restriction of the conformation, definition of a pharmacophore, and discovery of a peptide mimetic can be guided to some extent by molecular modeling. The scope and limitations of this tool will appear even more clearly in the following examples.

8.5 Selected Examples

8.5.1 Substance P, Neurokinins, Bradykinin, and Protein G-Coupled Receptors

A considerable effort has been made, since the design of spantide I (Folkers et al. 1984), towards the discovery of peptide and nonpeptide antagonists of the neurokinins (review: Regoli et al. 1994), due to both the therapeutic potential of this class of compounds and their usefulness as pharmacological tools. While the size and structure of spantides I and II (Folkers et al. 1990) are clearly related to the natural peptide ligands (substance P, neurokinin A, neurokinin B), a number of small pseudopeptide and nonpeptide antagonists have been discovered which fulfill the goals of high potency, selectivity, and stability required for a potential therapeutic use. Rational peptide optimization has led to several short pseudopeptide analogs such as S16474 (Kucharczyk et al. 1993), FR113680 (Morimoto et al. 1992), or FK888 (Fujii et al. 1992). In these studies and also in the design of the NK2-selective tachykinin antagonist PD 147714 (Boyle et al. 1994), only scarce use of molecular modeling was made. In contrast, for a hexapeptide series of neurokinin receptor agonists, a convincing conformational study helped to define the necessary geometry of the central core to confer NK1 or NK2 receptor selectivity (Deal et al. 1992).

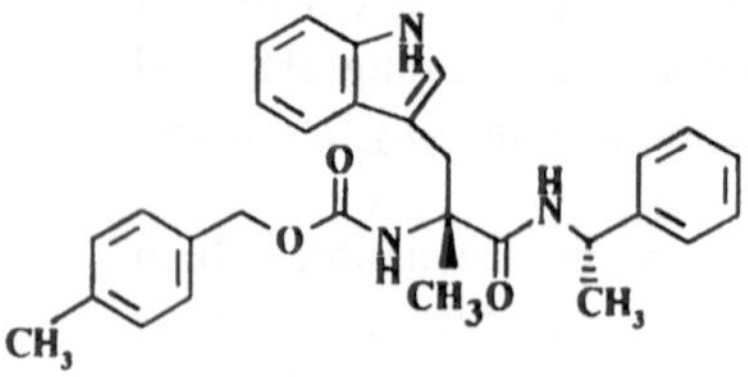

Cam 2445
NK_1, IC_{50} = 17nM

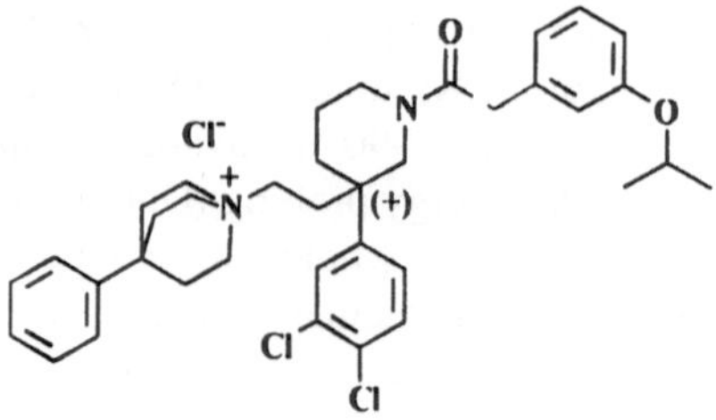

SR 140333
NK_1, pA_2 9 8

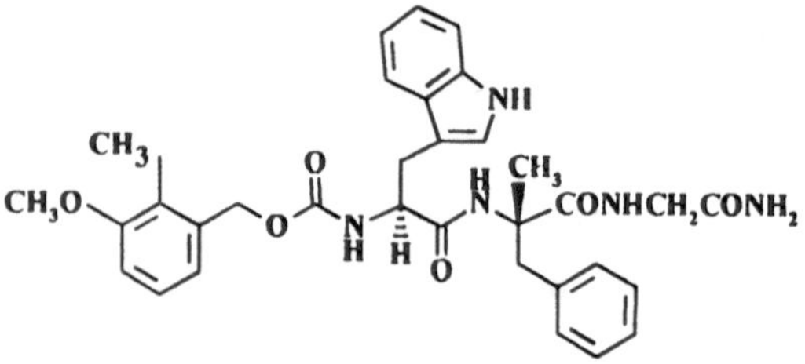

Cam 2291
NK_2, IC_{50} = 2nM

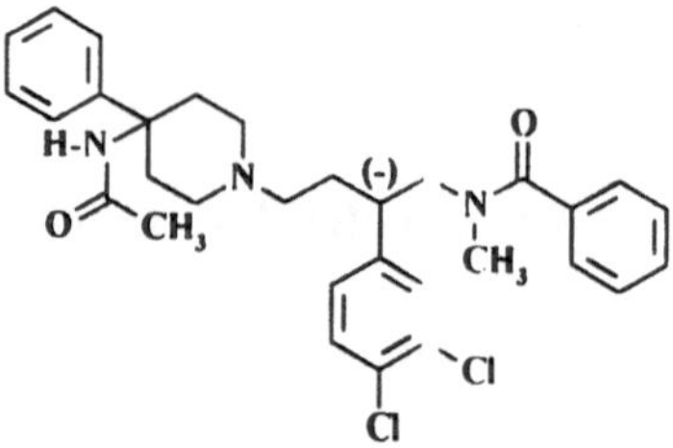

SR 48968
NK_2, pA_2 9.7

Boyle et al. 1994   **Emonds-Alt et al. 1993**

Fig. 1. Pairs of selective NK1/Nk2 receptor antagonists

One reason for the role of molecular modeling being moderate in the initial discovery step was that many nonpeptide antagonists were found via extensive screening of large compound collections, followed later by (nonpeptide) lead optimization. However, a large number of molecular modeling studies, mostly unpublished, were started by the discovery of the first nonpeptide NK1 antagonist CP96345 (Snider et al. 1991) and amplified by further discoveries (overview: Watling 1992), with the goal of identifying the common features of these new structures and of the natural peptide ligands or the peptide antagonists spantide (undecapeptide), S16474 (tetrapeptide), FR113680 (tripeptide), or FK888 (dipeptide). Superimpositions of the minimized conformations of these ligands apparently reveal strikingly common requirements for activity.

As a matter of fact, excellent 3D fitting is obtained of peptide and nonpeptide antagonists, on the one hand, and of the C-terminus of the neurokinins with all other ligands, on the other (see Regoli et al. 1994), which makes probable a common binding mode and binding site of these apparently quite different structures.

The question of selectivity towards one of the three known receptor types NK1, NK2, or NK3 has also been examined using molecular modeling. While, for example, obvious structural differences between the two pairs of NK1/NK2 antagonists are detectable in (Fig. 1: CAM 2445 and CAM 2291 (Boyle et al. 1994) and SR 140333 and SR 48968 (Emonds-Alt et al. 1993), it is questionable whether these models would permit one to design two related ligands with predicted NK1 and NK2 selectivity. The situation is even worse for the small number of known NK3 ligands, for which no clear-cut 3D features can presently be defined.

A more general problem in this field was raised by the discovery of different binding modes of spantide (as a representative relatively long peptide ligand), on the one hand, or of the nonpeptide NK1 ligand L703606 (iodinated analog of CP96345) or NK2 ligand SR48968, on the other. It was shown by Fong et al. (1992) and Gether et al. (1993) in elegant studies of chimeric NK1/NK2 or NK1/NK3 receptors that several receptor loops are involved in the interaction with substance P, neurokinin A, or spantide. Small nonpeptide antagonists act through short, discontinuous epitopes, which in the case of NK1 sites are located at the top of transmembrane fragments TM V and TM VI. This view is fully in line with modeling studies of the G protein-coupled neurokinin receptors (Trumpp-Kallmeyer et al. 1994) and of their interaction with substance P. Hence, rather than being simple competitive mimetics, these small antagonists allosterically disturb the interaction of the TM fragments with the G protein and prevent the formation of a high-affinity binding of the natural agonists (Gether et al. 1993). These facts raise the question of whether superimposition of the bioactive ligands and search of a common pharmacophore in molecular modeling is relevant (as it also raises the question of the relevance of competitive binding studies!).

In the terms of the message/address model of peptide receptor interaction (Schwyzer 1973), TM I to TM IV domains would recognize the address of the agonist and strengthen its binding, while TM V to TM VII would recognize the message and trigger the receptor/G protein interaction. Peptide antagonists, such as spantide, would mainly bind to the address-recognizing part of the receptor, thus fulfilling the proper conditions of competitive binding with respect to substance P (Trumpp-Kallmeyer et al. 1994). In contrast, small, generally nonpeptide antagonists, such as CP96345, most probably interact directly with the message site, thus exerting a negative intrinsic activity, according to Schütz and Freissmuth (1992).

More generally, site-directed mutagenesis studies of the G protein-coupled, seven transmembrane domain receptor family as well as elegant molecular modeling studies (review: Hibert et al. 1993) have established that antagonists often do not bind to the same site on the receptor as the peptide agonist. It should be stressed that the type of binding is in no way bound to the peptide nature of the ligand. Properties such as size, shape, and hydrophobicity obviously determine whether a peptide antagonist binds competitively (like spantide), or allosterically (like CP96345 or SR48968), after partial penetration in a binding pocket between defined transmembrane helices. Most likely, the peptide-derived FR113680, S16474, and FK 888 act in the second way, as suggested by molecular modeling.

For these reasons, the definition of a common pharmacophore for any of the neurokinin receptors, as attempted via superimpositions of known potent and selective antagonists (see above), may be futile and certainly cannot rule out the possibility of entirely unrelated, yet highly effective, configurations, as might be discovered by broad screening.

Very similar discoveries are currently being made with the development of bradykinin B2 antagonists (Kyle and Burch 1993). The first potent antagonists, such as NPC 17331 (Kyle et al. 1991), HOE 140 (Hock et al. 1991; Wirth et al. 1991), and S16118 (Thurieau et al. 1994a), in the size of nona- to decapeptides, have been designed from bradykinin by modification of the side chains and conformational restriction of the C-terminus. Extensive molecular modeling (Kyle et al. 1993) has convincingly demonstrated the presence of a β-turn involving the four C-terminal residues with a clear hydrogen bonded Arg9-NH. Furthermore, a model for the binding of bradykinin to its receptor has been proposed (Kyle et al. 1993), partly on the basis of site-directed mutagenesis, and partly on the modeling results of G-protein coupled receptors (Hibert et al. 1991). While the conclusions are that the above antagonists (which are the same size as bradykinin) bind in a competitive manner to the bradykinin site, it can be expected that shorter cyclic peptide (encompassing the C-terminus) or nonpeptide antagonists will be found, endowed with "intrinsic" antagonism (Schütz and Freissmuth 1992). N-terminally cyclized analogs (Chakravarty et al. 1993) and a mimic of NPC 17731 in which a stretch of the N-terminus was replaced by a 12-carbon chain (Kyle et al. 1994) displayed low pA_2 values.

8.5.2 RGD Pseudopeptides and Mimics

The tripeptide motif Arg-Gly-Asp (RGD) has been demonstrated to be a widespread recognizing signal for a number of adhesion proteins (Pierschbacher and Ruoslahti 1984). Considering that it is found twice within the Aα chain of fibrinogen, it was soon realized that peptides derived from the free tripeptide compete with fibrinogen for the binding to the platelet glycoprotein GPIIb/IIIa and inhibit platelet aggregation. A tremendous effort was made to optimize both peptide and nonpeptide RGD-like ligands as potential antithrombotic agents (reviews: Nichols et al. 1992; Gould 1993). The small size of the conserved segment promoted the synthesis of a large number of RGD-containing oligopeptides from which it could be read, for example, that RGDW was ten times more active than RGDS (Charon et al. 1990) and 4-piperidylcarbonyl-RGDW ten times more active than RGDW (Fauchère et al. 1993). It was also found that RGE was not recognized by integrins (D'Souza et al. 1991), while the replacement of arginine by lysine strongly increased the selectivity towards GPIIb/IIIa (Scarborough et al. 1993).

Although small peptide segments, in the size of tetra- to pentapeptides, can prevent platelet aggregation, it must be kept in mind that RGD (or KGD) is originally inserted in a protein chain and that specific recognition of this signal probably mainly involves the side chains of arginine and aspartic acid. As a matter of fact, further SAR studies with synthetic ligands demonstrated the outmost importance of the distance between the positive and negative charges of the side chain of these two residues: estimated range of intercharge distance 11 ± 3 Å. (Rao 1992; Greenspoon et al. 1993). Highly potent linear molecules were found

Table 1. Optimization of RGD analogs and mimics

Progress	Methods
RGD motif in integrins : a common cell recognition determinant	Prediction of turns (Rose 1978), of secondary structure (Chou and Fasman 1977), synthetic peptides
Optimization of linear peptides	Synthetic peptides/amino acid - substitutions/SAR
Solution conformation of linear oligopeptides: evidence for nested β-bends in GRGDSP	NMR in water and DMSO, CD
Vitronectin receptor selective cyclic analog	Cyclization and restriction of conformation, NMR
Conformation from X-ray of three main conformational types, presence of turns	X-ray analysis, sequence and structure comparison
GPIIb/IIIa selective cyclic analogs[a] SFK 106760 (PRP/ADP)[b] 0.36 μM c[RGDdFV] and c(RGDFdV) Ac-c[Pen/Cys-RGD-Cys)	Cyclization, NMR, X-ray Same lead sequence, different chirality Comparison Pen-Cys and Cys-Cys cyclization, NMR Chiral sulfoxides, NMR
Cyclo-S-acetyl-peptide, increased rigidity: 0.15 μM SKF 107260: 0.04 μM	Mercaptobenzoyl, mercaptoaniline surrogates of Cys
Demethylated SKF 107260	
Optimized linear pseudopeptides[a] RO-435054, RO-44 9883 SC-54701/ SC-54684	Synthetic analogs, SAR, random screening Acid/ester pair, prodrug
Optimized cyclic pseudopeptides[a] COR-72 (PRP/ADP): 0.5 μM TP 9201: 0.22 μM MK-0852: 0.1 μM DMP 728: 0.05 μM	SAR: variable cycle size and amino acid composition of cyclic pseudopeptides
Cyclic peptides, increased rigidity	Variable linkage in cylic analogs
Nonpeptides (peptide mimics) [a] L709780	Data base screening, SAR, centrally constrained mimics
SB 207448 / 208651	Benzodiazepine scaffold, N-methylation

Table 1 (continued)

Modeling and computational methods	References
Hydrophobicity profiles (Kyte and Doolittle 1982)	Pierschbacher and Ruoslahti 1984
Measurement of interchange distance, first superimpositions	Gartner et al. 1987; Charon et al. 1990; Samanen et al. 1991; Scarborough et al. 1993
Display of NMR structures (Havel and Wüthrich 1984), distance geometry, restrained molecular dynamics, energy minimization	Reed et al. 1988
Asp^- Arg^+ salt bridge revealed by constrained MD simulation	Siahaan et al. 1990
Display of X-ray structures, Monte Carlo and molecular dynamics of isolated RGDS	Cotrait et al. 1992
Constrained distance geometry agreement of structures in solution and crystal	Kopple et al. 1992
Molecular mechanics and dynamics to explain laminin/vitronectin ratio	Müller et al. 1992
Modeling and distance geometry	Bogusky et al. 1993
Molecular dynamics of rigid isomer template	McDowell and Gadek 1992
Molecular dynamics and genetic algorithm to fit NMR and CD data	Ali et al. 1994; Sanderson et al. 1994
None reported, induces conformational changes on GP IIb/IIIa	Alig et al. 1992; Kouns et al. 1992; Zablocki et at 1994
Superimposition	Review: Gould 1993
Molecular dynamics	Cheng et al. 1994
Molecular dynamics: consensus conformation	Mc Dowell et al. 1994a
Modeling: exclusion of conformation of weakly active analogs	Egbertson et al. 1994a,b
Modeling, NMR. X-ray	Ku et al. 1993; McDowell et al. 1994b

Pen, penicillamine. [a]For structure formula, see Gould 1993 (review) and original references. [b] IC_{50} for ADP-induced platelet aggregation in platelet-rich plasma.

which inhibit aggregation of human plasma with IC_{50} values of 40–50 μM (Kouns et al. 1992; Zablocki et al. 1993).

The complementary observation that conformation of the tripeptide strongly modulated activity was also believed to be mediated by a modulation of the intercharge distance (Rao 1992). Potent cyclic pseudopeptides were then developed in order to restrict the conformational freedom and to increase potency and selectivity. While the first cyclic antagonist was vitronectin receptor selective (Siahaan et al. 1990), efforts concentrated mostly towards GPIIb/IIIa-specific ligands (Samanen et al. 1991; Nutt et al. 1992). These mainly cyclic disulfides were extremely useful to define the 3D requirements for the RGD segment. However, most of these analogs were not active if orally administered. Since the real challenge was to design orally active compounds, the cyclic templates were used to suggest nonpeptide mimetics, which hopefully would promote oral absorption. Some of the resulting compounds (Table 1) display impressive IC_{50} values in vitro and several analogs are orally active as such, while others have to be prepared as prodrugs.

Molecular modeling played an important role in several of the key steps outlined above. Firstly, it made possible to analyze the X-ray crystal structures from the Brookhaven protein data bank and to classify the conformations of the RGD motif into only three main structural patterns, according to the intercharge distance δ of the Arg and Asp side chains (Cotrait et al. 1992; Rao 1992). It also allowed the structure with $\delta \geq 10$ Å to be identified as the most probable one (for GPIIB/IIIa selectivity) on the basis of energy calculations incorporating the solvation in relation to the solvent-accessible surface area (Mohamadi et al. 1990). Furthermore, molecular modeling was certainly involved in the development of non cyclic mimics of RGDX (X = any amino acid) via superimposition of either the tested or the planned analogs (see centrally constrained inhibitors: Egbertson et al. 1994b). The identification of turns in linear RGD-containing segments by modeling (and by NMR or X-ray translated into manageable graphics) helped in the design of cyclic peptides since forced turns are the first step toward cyclization. Modeling confirmed the strategies known to reduce conformational flexibility such as N-methylation and cyclization (Samanen et al. 1991) and suggested the sequence (incorporation of additional residues) in order to conserve the original turn. Modeling and molecular dynamic analysis helped to detect a consensus conformation in these cyclic analogs (McDowell et al. 1994a), which have become major templates for the discovery of nonpeptide mimetics. Modeling also guided the replacement of the peptide cycle by a convenient scaffold (benzodiazepine: Ku et al. 1993, McDowell et al. 1994b; tyrosine: Egbertson et al. 1994a; isoindolinone: Egbertson et al. 1994b) with a similar 3D presentation of the binding determinants. A concomitant increase of the oral bioavailability compared to the original peptides has been observed in the benzodiazepine series (McDowell et al. 1994b) but also some encouraging results of pharmacokinetic studies with optimized linear peptide-derived compounds, administered orally, without the prodrug modification (e.g., L 703714: Barrett et al. 1994).

On the whole, the sequence RGD was the object of intense work in which molecular modeling certainly was involved (Table 1 and review: Mager 1994). However, the potent and stable derivatives discovered often still require the prodrug approach to be correctly absorbed and will have to face as drugs the competition with the recently approved anti-GPIIb/IIIa antibody (Simoons et al. 1994). In contrast, a promising application in wound healing of RGD-containing segments (linear polycationic octadecapeptide covalently bound to hyaluronic acid) has been described (Polarek et al. 1994).

8.5.3 Inhibitors of Proteases and Structure-Based Design

Structure-based (drug) design (SBDD) is a fundamentally distinct approach which requires knowing the X-ray structure of the macromolecular target or ideally of the cocrystallized enzyme–inhibitor or receptor–ligand complex (reviews: Reich and Webber 1993; Greer et al. 1994). Molecular graphics are then used as the basis for docking potential ligands into the active site. Identification and optimization of peptide leads in this area have focused mainly on protease inhibitors. Angiotensin-inhibiting enzyme (ACE) inhibitors, historically the first major pharmaceutical targets in this field, were not discovered by SBDD (no crystals available!) but rather by SAR studies on existing peptide substrates (review: Lawton et al. 1992). Renin inhibitors represent an intermediate stage of design in which both extensive SAR studies on prototype inhibitors (pepstatin) and docking experiments with available crystal structures of aspartyl proteases and, finally, of human renin (Sibanda et al. 1984) were performed (review: Greenlee 1990; Humblet et al. 1993).

HIV protease inhibitors are the first cases where most of the design was based on the crystal structure, taking in account the homodimeric composition of the aspartyl protease with each monomer contributing one aspartic acid to the active site. Either pepstatin-based, symmetric, or nonpeptide constrained inhibitors were designed as complementary to the binding cleft and a large number of protein-inhibitor crystal structures were solved during the design process as a guide for further syntheses (Wlodawer and Erickson 1993). Since SBDD does not necessarily require knowledge of preexisting ligands, de novo synthesis of truly new ligands is possible [see software GROW (Moon and Howe 1991) and LUDI (Böhm 1992)]. HIV protease inhibitors have also benefited from this approach as seen in the design of DMP323, in which a cyclic urea carbonyl both mimics and displaces a structural water molecule in the enzyme–inhibitor complex (Lam et al. 1994).

In view of the several optimized compounds known to be undergoing clinical trials, such as Ro 318959 (Krohn et al. 1991), A77003 (Kempf et al. 1993), S52151 (Getman et al. 1993), DMP323 (Lam et al. 1994), and AG 1284 (Reich 1994), the feeling is that this field matured over a shorter period of time than did renin inhibitors mainly due to the knowledge of X-ray structures and to the corresponding opportunities for molecular modeling.

8.5.4 What Brought About the Breakthrough?

Among the peptides approved for human use (Verlander et al. 1991), all levels of sophistication are encountered, from the natural unmodified sequence, over the pseudopeptide, to the nonpeptide mimetic of the original peptide. A number of synthetic peptides are unmodified [salmon calcitonin (Azria 1989), cyclosporin (Wenger 1984), thymopentin (Malaise et al. 1985)] or derivatized [pentagastrin (Morley et al. 1965), synacthen (Rittel 1973)]. The five following major peptide drugs, either already on the US market or expected by 1996, are optimized peptide derivatives for which it is recognized that molecular modeling only rarely played a decisive role. Hence, desmopressin was the result of a molecular tailoring (deamination of hemicystine-1 and D-Arg substitution for Arg in position 8) which increased the desired antidiuretic effect and decreased the pressor and uterotonic effects (review: Richardson and Robinson 1985). Similarly, the conversion of luteinizing hormone–releasing hormone (LHRH; GnRH) agonists into effective and manageable drugs against prostate cancer came as much from the incorporation of the peptide into a biodegradable subcutaneous implant (Perren et al. 1986) as from the chemical optimization of the hormone (Dutta et al. 1978). The spectacular reduction in size of somatostatin obtained with the potent analogs MK678 (review: Veber 1992) and octreotide (Bauer et al. 1982) also relied mainly on NMR investigations and on stabilization towards proteolysis and only moderately on molecular modeling: the development of the second of these minisomatostatins depended on the possible minimization of an adverse effect: steatorrhea! (Hirschmann 1991). The bivalent synthetic hirudin analogs could probably not have been designed without knowledge of the crystal structure of the hirudin/α-thrombin complex (Grütter et al. 1990) which (on molecular graphics) helped to estimate the distance between the groove on the protein surface and the active site, thus leading to the second generation of inhibitors hirulog (Maraganore et al. 1990) and hirofos (Thurieau et al. 1994b), compared to the less potent C-terminal fragments (MDL28050: Krstenansky et al. 1990). Finally, the RGD peptides, in contrast to their mimetic derivatives, will depend on the galenic preparation for their successful application in wound healing (Polarek et al. 1994).

Similar observations can be made for a few other peptide-derived drug candidates. Even if the docking of ligands into the active site of renin was instrumental for the design of potent inhibitors (review: Humblet et al. 1993), the way to oral inhibitors was opened by monitoring the blood plasma levels of the analogs after oral administration and simultaneous optimization of potency and absorption (Rosenberg et al. 1993). The optimization of the potent VIP agonist Ro 25 1553 (Bolin et al. 1994) was based mainly on the concomitant cyclization and stabilization against proteolysis of the analogs of the native peptide, without strong reference to modeling studies. Detailed 3D elements which helped to progressively reduce the size and restrict the conformational freedom of neuropeptide Y (NPY) came first from computer-aided comparison of NPY with the X-ray structure of pancreatic polypeptide which belongs to the same structural family (Allen et al. 1987). Spectral studies and molecular dynamic simulations revealed a defined poly-proline-type II helix in residues 1–8, a β-turn through positions 9–14, an amphipathic α-helix between residues 15 and 32 and C-terminal turn structure. Although molecular modeling constantly guided these studies [using mainly the consistent force field of Hagler et al. (1974) for dynamics]), it was, as in the largest number of cases, just one of the several tools which led to success (Kirby et al. 1993).

Finally, rare examples have been reported which point to molecular modeling and computer-aided design as the decisive tools for the discovery of the new drug. Cilazapril is one such example where a nonpeptide heterocyclic ring was clearly suggested by modeling and where modeling studies helped to define the appropriate ring size and the best positions for the substituents (review: Hangauer 1989). This modeling study skillfully exploited the large body of structural information available from previous work on peptide-derived ACE inhibitors. However, external information or chemical intuition is generally required as well. Hence, some of the linear nonpeptide RGD-like GPIIb/IIIa receptor antagonists such as SB207448 and its N-methylated derivative (Ku et al. 1993) which are claimed to have been discovered on the basis of 3D-matching to the selective and potent cyclic RGD-containing peptide templates have required first the traditional optimization of these templates, the structural information available from NMR and X-ray, followed by an arbitrary choice of the central (benzodiazepine) template, in addition to molecular modeling.

8.6 Conclusion/Outlook

The attractive and powerful concept of the complementary lock and key (Fischer 1894) had to undergo considerable refinement in the last few years. In its most simple form, it may hold for protease inhibitors (and other enzyme ligands), except for some dynamically induced fit of the ligand into the catalytic site (deforming and nonrigid). It may even help to rationalize the interaction of the RGD-like ligands to the GPIIb/IIIa (and other integrins) receptors. In contrast, most of the peptide hormone agonists and antagonists act on a more complex structure, the seven transmembrane G-protein-coupled receptor. This lock, far from being a single cavity with a defined imprint, extends both on the cell surface and membrane (i.e., over domains of higly variable hydrophobicity) and can be distorted in such a way as to abolish coupling to the G-protein. It allows for different binding modes of agonist and antagonist and for multiple binding modes of different antagonists. While the endogenous agonist likely binds in a unique way, optimization of synthetic ligands, especially antagonists, may well end up into structurally unrelated compounds, all potent and selective in vitro. The final selection of therapeutic agents will be based on the achieved structure simplification (potency/molecular weight ratio), the metabolic stability and oral bioavailability, and the in vivo potency. Since the nondenaturing purification and crystallization of G protein-coupled receptors is still a major problem, molecular modeling is presently a convenient way to get insight into their 3D structure and, when associated to site-directed mutagenesis, a powerful tool to guide the optimization of peptide-derived ligands.

References

Ali FE, Bennett DB, Clavo RR, Elliott JD, Hwang SM, Ku TW, Lago MA, Nichols AJ, Romoff TT, Shah DH, Vasko JA, Wong AS, Yellin TO, Yuan CK, Samanen JM (1994) Conformationally constrained peptides and semipeptides derived from RGD as potent inhibitors of the platelet fibrinogen receptor and platelet aggregation. J Med Chem 37:769–780

Alig L, Edenhofer A, Hadvary P, H.rzeler M, Knopp D, Müller M, Steiner B, Trzeciak A, Weller T (1992) Low molecular weight, non peptide fibrinogen receptor antagonists. J Med Chem 35:4393–4407

Allen J, Novotny J, Martin J, Heinrich G (1987) Molecular structure of neuropeptide Y: analysis by molecular cloning and computer-aided comparison with crystal structure of avian homologue. Proc Natl Acad Sci USA 84:2532–2536

Attwood MR, Francis RJ, Hassall CH, Krohn A, Lawton G, Natoff IL, Nixon JS, Redshaw S, Thomas WA (1984) New potent inhibitors of angiotensin-converting enzyme. FEBS Lett 165:201–205

Audry E, Dubost JP, Dallet P, Langlois MH, Colleter JC (1989) Le potentiel de lipophilie moléculaire: application à une série d'amines β-adrénolytiques. Eur J Med Chem 24:155-161

Audry E, Dubost JP, Langlois MH, Croizet F, Braquet P, Dallet P, Colleter JC (1992) Use of molecular lipophilicity potential in QSAR. In: Kuchar M (ed)

QSAR in design of bioactive compounds, Prous Science, Barcelona, pp 249–268

Azria M (1989) The calcitonins. In: Azria M (ed) The calcitonins. Karger, Basel, pp 1–152

Baniak EL, Rivier JE, Struthers RS, Hagler AT, Gierasch LM (1987) Nuclear magnetic resonance analysis and conformational characterization of a cyclic decapeptide antagonist of GnRH. Biochemistry 26:2642–2656

Barra D, Mignogna G, Simmaco M, Pucci P, Severini C, Falconieri-Erspamer G, Negri L, Erspamer V (1994) [D-Leu2]deltorphin, a 17 amino acid opioid peptide from the skin of the brazilian hylid frog, phyllomedusa burmeisteri. Peptides 15:199–202

Barrett JS, Gould RJ, Ellis JD, Holahan MM, Stranieri MT, Lynch JJ, Hartman GD, Ihle NB, Duggan M, Moreno OA, Theoharides AD (1994) Pharmacokinetics and pharmacodynamics of L 703014, a potent fibrinogen receptor antagonist, after intravenous and oral administration in the dog. Pharm Res 11:426–431

Bauer W, Briner U, Doepfner W, Haller R, Huguenin R, Marbach P, Petcher TJ, Pless J (1982) SMS 201 995: a very potent and selective octapeptide analogue of somatostatin with prolonged action. Life Sci 31:1133–1140

Bogusky MJ, Naylor AM, Mertzman ME, Pitzenberger SM, Nutt RF, Brady SF, Colton CD, Veber DF (1993) The solution conformation of Ac-Pen-Arg-Gly-Asp-Cys-OH, a potent fibrinogen receptor antagonist. Biopolymers 33:1287–1297

Böhm HJ Jr (1992) LUDI: rule-based automatic design of new substituents for enzyme inhibitor leads. Comput Aided Mol Des 6:593–606

Bolin DR, Cottrell JM, Michalewsky J, Garippa R, Rinaldi N, O'Donnell M, Selig W (1994) Ro 25–1553: a potent, metabolically stable vasoactive intestinal peptide agonist. In: Hodges RS, Smith JA (eds) Peptides, chemistry, structure and biology. ESCOM, Leiden, pp 843–845

Boyd DB (1993) Compendium of software for molecular modeling. In: Lipkowitz KB, Boyd DB (eds) Reviews in computational chemistry, vol IV. VCH, New York, pp 229-257

Boyd DB, Lipkowitz KB (1982) Molecular mechanics. The method and its underlying philosophy. J Chem Educ 59:269–274

Boyle S, Guard S, Hodgson J, Horwell DC, Howson W, Hughes J, McKnight A, Martin K, Pritchard MC, Watling KJ, Woodruff GN (1994) Rational design of high affinity tachykinin NK2 receptor antagonists. Bioorg Med Chem 2:101–113

Buckert U, Allinger NL (1982) Molecular mechanics. American Chemical Society, Washington DC

Chakravarty S, Wilkins D, Kyle DJ (1993) Design of potent cyclic peptide bradykinin receptor antagonists from conformationally constrained linear peptides. J Med Chem 36:2569–2571

Charon MH, Poggi A, Donati MB, Marguerie G (1990) Synthetic peptides with antithrombotic activity. In: Rivier JE, Marshall GR (eds) Peptides, chemistry, structure and biology. ESCOM, Leiden, pp 82–83

Cheng S, Craig WS, Mullen D, Tschopp JF, Dixon D, Pierschbacher MD (1994) Design and synthesis of novel cyclic RGD containing peptides as highly potent and selective integrin αIIb/β3 antagonists. J Med Chem 37:1–8

Chou PY, Fasman GD (1977) Secondary structural prediction of proteins from their amino acid sequence. TIBS 2:128–131

Clark M, Cramer RD, Jones DM, Patterson DE, Simeroth PE (1990) Comparative molecular field analysis (CoMFA). 2. Toward its use with 3D-structural databases. Tetra Com Methods 3:47–59

Clore GM, Gronenborn AM, Carlson G, Meyer EF (1986) Stereochemistry of binding of the tetrapeptide Ac-Pro-Ala- Pro-Tyr-NH2 to porcine pancreatic elastase. Combined use of 2D-transferred nuclear Overhauser enhancement measurements, restrained molecular dynamics, X-ray crystallography and molecular modelling. J Mol Biol 190:259–267

Connoly ML (1983) Analytical molecular surface calculation. J Appl Crystallogr 16:548–558

Cotrait M, Kreissler M, Hoflack J, Lehn JM, Maigret B (1992) Computational simulations of the conformational behaviour of the adhesive proteins RGDS fragment. J Comput Aided Mol Des 6:113–130

Cramer RD (1993) Partial Least Squares (PLS): its strengths and limitations. Perspect Drug Discov Des 1:269–278

Cramer RD, Patterson DE, Bunce JD (1988) Comparative molecular field analysis (CoMFA). 1. Effect of shape on binding of steroids to carrier proteins. J Am Chem Soc 110:5959–5967

Crippen G (1981) Distance geometry and conformational calculations. In: Bawden D (ed) Chemometrics research studies series, vol 1. Wiley, New York

Deal MJ, Hagan RM, Ireland SJ, Jordan CC, McElroy AB, Porter B, Ross BC, Stephens-Smith M, Ward P (1992) Conformationally constrained tachykinin analogues: potent and highly selective NK2 receptor agonists. J Med Chem 35:4195–4204

D'Souza SE, Ginsberg MH, Matsueda GR, Plow EF (1991) A discrete sequence in a platelet integrin is involved in ligand recognition. Nature 350:66–68

Dutta AS, Furr BJ, Giles MB, Valcaccia B (1978) Synthesis and biological activity of highly active α-aza-analogues of luliberin. J Med Chem 21:1018–1024

Egbertson MS, Chang CT, Duggan ME, Gould RJ, Halczenko W, Hartman GD, Laswell WL, Lynch JJ, Lynch RJ, Manno PD, Naylor AM, Prugh JD,

Ramjit DR, Sitko GR, Smith RS, Turchi LM, Zhang G (1994a) Non-peptide fibrinogen receptor antagonists. Optimization of a tyrosine template as a mimic for Arg-Gly-Asp. J Med Chem 37:2537–2551

Egbertson MS, Naylor AM, Hartman GD, Cook JJ, Gould RJ, Holahan MA, Lynch JJ, Lynch MT, Stranieri MT, Vassallo LM (1994b) Non-peptide fibrinogen receptor antagonists. Design and discovery of a centrally constrained inhibitor. Bioorg Med Chem Lett 4:1835–1846

Emonds-Alt X, Doutremepuich JD, Heaulme M, Neliat G, Santucci V, Steinberg R, Vilain P, Bichon D, Ducoux JP, Proietto V, VanBroek D, Sobrié P, LeFur G, Brelière JC (1993) In vitro and in vivo biological activities of SR140333, a novel non-peptide tachykinin NK-1 receptor antagonist. Eur J Pharmacol 250:403–413

Erspamer V, Melchiorri P (1980) Active polypeptides: from amphibian skin to gastrointestinal tract and brain of mammals. TIPS 1:391–395

Fauchère JL (1984) QSAR of oligopeptides: amino acid side chain parameters and some specific studies. In: Kuchar M (ed) QSAR in design of bioactive compounds. Prous, Barcelona, pp 135–144

Fauchère JL (1986) Elements for the rational design of peptide drugs. Adv Drug Res 15:29–69

Fauchère JL (1989) Towards the rational design of peptide drugs. Actual Chim Therap 16:55–72

Fauchère JL (1994) Peptides: multiple purpose tools. In: Danielsson B, Birnbaum S, Bülow L, Larsson PO, Mansson MO (eds) Advances in molecular and cell biology, biochemical technology, vol 2. Jai Press, Greenwich, pp 83–99

Fauchère JL, Lauterwein J (1985) The chemical shift of the alpha carbon in amino acids as a parameter for QSAR studies of oligopeptides. Quant Struct Act Relat 4:13–18

Fauchère JL, Thurieau C (1992) Evaluation of the stability of peptides and pseudopeptides as a tool in peptide drug design. Adv Drug Design 23:127–159

Fauchère JL, Quarendon P, Kaetterer L (1988) Estimating and representing hydrophobicity potential. J Mol Graph 6:202–206

Fauchère JL, Morris AD, Thurieau C, Simonet S, Verbeuren TJ, Kieffer N (1993) Modulation of the activity and assessment of the receptor selectivity in a series of new RGD-containing peptides. Int J Peptide Protein Res 42:440–444

Fesik SW, Gampe RT, Eaton HL, Gemmecker G, Olejniczak ET, Neri P, Holzman TF, Egan DA, Edalji R, Simmer R, Helfrich R, Hochlowski J, Jackson M (1991) NMR studies of [U-13C]cyclosporin A bound to cyclophilin: bound conformation and portions of cyclosporin involved in binding. Biochemistry 30:6574–6583

Fischer E (1894) Einfluss der Configuration auf die Wirkung der Enzyme. Chem Ber 27:2985–2993

Folkers K, Häkanson R, Hörig J, Xu JC, Leander S (1984) Biological evaluation of substance P antagonists. Br J Pharmacol 83:449–456

Folkers K, Feng DM, Asano N, H.kanson R, Weisenfeld-Hallin Z, Leander S (1990) Spantide II, an effective tachykinin antagonist having high potency and negligible neurotoxicity. Proc Natl Acad. Sci USA 87:4833–4835

Fong TM, Huang RRC, Strader C (1992) Localization of agonist and antagonist binding domains of the human neurokinin-1 receptor. J Biol Chem 267:25664–25667

Fournié-Zaluski MC, Belleney J, Lux B, Durieux C, Gerard D, Gacel G, Maigret B, Roques BP (1986) Conformational analysis of CCK26-33 and related fragments by 1H NMR spectroscopy, fluorescence transfer measurements and calculations. Biochemistry 25:3778–3787

Fujii T, Murai M, Morimoto H, Maeda Y, Yamaoka M, Hagiwara D, Miyake H, Ikari N,Matsuo M (1992) Pharmacological profile of a high affinity dipeptide NK-1 receptor antagonist, FK888. Br J Pharmacol 107:785–789

Furet P, Sele A, Cohen NC (1988) 3D molecular lipophilicity potential profiles: a new tool in molecular modeling. J Mol Graph 6:182–200

Gaillard P, Carrupt PA, Testa B, Boudon A (1994) Molecular lipophilicity potential, a tool in 3d QSAR: methods and applications. J Comput Aided Mol Des 8:83–96

Gallop MA, Barrett RW, Dower WJ, Fodor SP, Gordon EM (1994) Applications of combinatorial technologies to drug discovery. 1. Background and peptide combinatorial libraries. J Med Chem 37:1233–1251

Gartner TK, Bennett JS, Olgivie ML (1987) Effects of Ala for Gly substitutions in the GD regions of the peptides RGDS and LGGAKQAGDV. Blood 70 [S1]:351a

Gether U, Johansen TE, Snider RM, Lowe JA III, Emonds-Alt X, Yokota Y, Nakanishi S, Schwartz TW (1993) Binding epitopes for peptide and non peptide ligands on the NK1 (substance P) receptor. Regul Peptides 46:49–58

Getman DP, DeCrescenzo GA, Heints RM, Reed KL, Talley JJ, Bryant ML, Clare M, Houseman KA, Marr JJ, Mueller RA, Vazquez ML, Shieh HS, Stallings WC, Stegeman RA (1993) Discovery of a novel class of potent HIV-1 protease inhibitors containing the (R)-(hydroxyethyl)urea isostere. J Med Chem 36:288–291

Goodford PJ (1984) Drug design by the method of receptor fit. J Med Chem 27:557–564

Gordon EM, Barrett RW, Dower WJ, Fodor SP, Gallop MA (1994) Applications of combinatorial technologies to drug discovery. 2. Combinatorial organic synthesis, library screening strategies and future directions. J Med Chem 37:1385–1401

Gould RJ (1993) The integrin αIIbβ3 as an antithrombotic target. Perspect Drug Discov Des 1:537–548

Greenlee WJ (1990) Renin inhibitors. Med Res Rev 10:173–236 Greenspoon N, Hershkoviz R, Alon R, Varon D, Shenkman B, Marx G, Federman S, Kapustina G, Lider O (1993) Structural analysis of integrin recognition and the inhibition of integrin-mediated cell functions by novel nonpeptidic surrogates of the Arg-Gly-Asp sequence. Biochemistry 32:1001–1008

Greer J, Erickson JW, Baldwin JJ, Varney MD (1994) Application of the 3D structures of protein target molecules in structure-based drug design. J Med Chem 37:1035–1054

Griesinger C, Sorensen OW, Ernst RR (1987) Novel three-dimensional NMR techniques for studies of peptides and biological macromolecules. J Am Chem Soc 109:7227–7228

Grütter MG, Priestle JP, Rahuel J, Grossenbacher H, Bode W, Hofsteenge J, Stone SR (1990) Crystal structure of the thrombin-hirudin complex: a novel mode of serine protease inhibition. EMBO J 9:2361–2365

Hagler AT, Huler E, Lifson S (1974) Energy functions for peptides and proteins: derivation of a consistent force field including the hydrogen bond from amide crystals. J Am Chem Soc 96:5319–5327

Hahn KW, Klis WA, Stewart JM (1990) Design and synthesis of peptide having chymotrypsin-like esterase activity. Science 248:1544–1547

Hangauer DG (1989) Computer-aided design and evaluation of angiotensin-converting enzyme inhibitors. In: Perun TJ, Propst CL (eds) Computer-aided drug design, Dekker, New York, pp 253–295

Hansch C, Leo A (1979) Substituent constants for correlation analysis in chemistry and biology. Wiley, New York, pp 13–43

Havel T, Wüthrich K (1984) A distance geometry program for determining the structures of small proteins and other macromolecules from NMR measurements of intramolecular 1H-1H proximities in solution. Bull Math Biol 46:673–698

Hellberg S, Sjöström M, Wold S (1986) The prediction of bradykinin potentiating potency of pentapeptides: an example of a peptide quantitative structure-activity relationship. Acta Chem Scand B40:135–140

Hibert MF, Trumpp-Kallmeyer S, Bruinvels A, Hoflack J (1991) Three-dimensional models of neurotransmitter G-binding protein-coupled receptors. Mol Pharmacol 40:8–15

Hibert MF, Trumpp-Kallmeyer S, Hoflack J, Bruinvels A (1993) This is not a G protein-coupled receptor. Trends Pharmacol Sci 14:7–12

Hirschmann R (1991) Medicinal chemistry in the golden age of biology: lessons from steroid and peptide research. Angew Chem Int Ed Engl 30:1278–1301

Ho CM, Marshall GR (1990) Cavity search: an algorithm for the isolation and display of cavity-like binding regions. J Comput Aided Mol Design 4:337–354

Hock FG, Wirth K, Albus U, Linz W, Gerhards HJ, Wiemer G, Henke S, Breipohl G, König W, Knolle J, Schölkens BA (1991) HOE140 a new potent and long-acting bradykinin antagonist: in vitro studies. Br J Pharmacol 102:769–773

Hol WG, Van Duijnen PT, Berendsen HJ (1978) The α-helix dipole and the properties of proteins Nature 273:443–446

Houghten R (1993) Peptide libraries: criteria and trends. Trends Genet 9:235–239

Humblet C, Lunney EA, Mirzadegan T (1993) Docking ligands in the receptor cavity: what have we learned? In: Wermuth CG (ed) Trends in QSAR and molecular modelling 92. ESCOM, Leiden, pp 35–43

Kellogg GE, Semus SF, Abraham DJ (1991) HINT: a new method of empirical hydrophobic field calculation for CoMFA. J Comput Aided Mol Des 5:545–552

Kempf DJ, Codacovi L, Wang XC, Kohlbrenner WE, Wideburg NE, Saldivar A, Vasavanonda S, Marsh KC, Bryant P, Sham HL, Green BE, Betebenner DA, Erickson J, Norbeck DW (1993) Symmetry-based inhibitors of HIV protease: structure-activity studies of acylated 2,4-diamino-1,5-diphenyl-3-hydroxypentane and 2,5-diamino-1,6-diphenylhexane-3,4-diol. J Med Chem 36:320–330

Kessler H, Bermel W, M.ller A, Pook KH (1985) Modern nuclear magnetic resonance spectroscopy of peptides. In: Gross E, Meienhofer J (eds) The peptides, vol 7. Academic, New York, pp 437–473

Kessler H, Klein M, Miller A, Wagner K, Bats JW, Ziegler K, Frimmer M (1986) Conformational prerequisites for the in vitro inhibition of cholate uptake in hepatocytes by cyclic analogues of antamanide and somatostatin. Angew Chem (Int Ed Engl) 25:997–999

Kirby DA, Koerber SC, Craig AG, Feinstein RD, Delmas L, Brown MR, Rivier JE (1993) Defining structural requirements for neuropeptide Y receptors using truncated and conformationally restricted analogues. J Med Chem 36:385–393

Kirkpatrick S, Gelatt CD, Vecchi MP (1983) Optimization by simulated annealing. Science 220:671–680

Koltun WL (1965) Precision space-filling atomic models. Biopolymers 3:665–669

Kopple KD, Baures PW, Bean JW, D'Ambrosio CA, Hughes JL, Peishof CE, Eggleston DS (1992) Conformations of Arg-Gly-Asp containing heterodetic cyclic peptides: solution and crystal studies. J Am Chem Soc 114:9615–9623

Kouns WC, Hadvary P, Hearing P, Steiner B (1992) Conformational modulation of purified glycoprotein GPIIb-IIIa allows proteolytic generation of active fragments from either active or inactive GPIIb-IIIa. J Biol Chem 267:18844–18851

Krohn A, Redshaw S, Ritchie JC, Graves BJ, Hatada MH (1991) Novel binding mode of highly potent HIV proteinase inhibitors incorporating the (R)-hydroxyethylamine isostere. J Med Chem 34:3340–3342

Krstenansky JL, Broersma RJ, Owen TJ, Payne MH, Yates MT, Mao SJ (1990) Development of MDL 28050, a small stable antithrombin agent based on a functional domain of the leech protein hirudin. Thromb Haemost 63:208–214

Ku TW, Ali FE, Barton LS, Bean JW, Bondinell WE, Burgess JL, Callahan JF, Calvo RR, Chen L, Eggleston DS, Gleason JG, Huffman WF, Hwang SM, Jakas DR, Karash CB, Keenan RM, Kopple KD, Miller WH, Newlander KA, Nichols A, Parker MF, Peishoff CE, Samanen JM, Uzinslas I, Venslavsky J (1993) Direct design of a potent non peptide fibrinogen receptor antagonist based on the structure and conformation of a highly constrained cyclic RGD peptide. J Am Chem 115:8861–8862.

Kubinyi H (1993) CoMFA. In: Mannhold R, Krogsgaard P, Timmerman H (eds) QSAR: Hansch analysis and related approaches, VCH, Weinheim, pp 159–172

Kucharczyk N, Thurieau C, Paladino J, Morris AD, Bonnet J, Canet E, Krause JE, Regoli D, Couture R, Fauchère JL (1993) Tetrapeptide tachykinin antagonists: synthesis and modulation of the physicochemical and pharmacological properties of a new series of partially cyclic analogs. J Med Chem 36:1654–1661

Kyle DJ, Burch RM (1993) A survey of bradykinin receptors and their antagonists. Curr Opin Invest Drugs 2:5–20

Kyle DJ, Martin JA, Burch RM, Carter JP, Lu S, Meeker S, Prosser JC, Sullivan JP, Togo J, Noronha-Blob L, Sinsko JA, Walteers RF, Whaley LW, Hiner RN (1991) Probing the bradykinin receptor: mapping the geometric topography using ethers of hydroxyproline in novel peptides. J Med Chem 34:2649–2653

Kyle DJ, Blake PR, Smithwick D, Green LM, Martin JA, Sinsko JA, Summers MF (1993) NMR and computational evidence that high-affinity bradykinin receptor antagonists adopt C-terminal β-turns. J Med Chem 36:1450–1460

Kyle DJ, Chakravarty S, Sinsko JA, Stormann TM (1994) A proposed model of bradykinin bound to the rat B2 receptor and its utility for drug design. J Med Chem 37:1347–1354

Kyte J, Doolittle RF (1982) A simple method for displaying the hydropathic character of a protein. J Mol Biol 157:105–132

Lam PY, Jadhav PK, Eyermann CJ, Hodge CN, Ru Y, Bacheler LT, Meek JL, Otto MJ, Rayner MM, Wong YN, Chang CH, Weber PC, Jackson DA,

Sharpe TR, Erickson-Vitanen S (1994) Rational design of potent, bioavailable, nonpeptide cyclic ureas as HIV protease inhibitors. Science 263:380–384

Lawton G, Paciorek PM, Waterfall JF (1992) The design and biological profile of ACE inhibitors. Adv Drug Res 23:161–220

Lipkowitz KB, Boyd DB (eds) (1991) Molecular mechanics methods. VCH, New York, pp 1–158 (Reviews in computational chemistry)

Lipton M, Still WC (1988) The multiple minimum problem in molecular modeling. Tree searching internal coordinate conformational space. J Comput Chem 9:343–355

Mager PP (1994) Interactive multivariate modeling of ArgGlyAsp (RGD) Derivatives. Med Res Rev 14:75–126

Malaise MG, Franchimont P, Bach-Andersen R, Gerber H, Stocker H, Hauwaert C, Danneskiold B, Gross D, Gerschpacher H, Bolla K (1985) Treatment of active rheumatoid arthritis with slow intravenous injections of thymopentin. Lancet 13:832–836

Mammi NJ, Hassan M, Goodman M (1985) Conformational analysis of a cyclic enkephalin analogue by 1H NMR and computer simulations. J Am Chem Soc 107:4008–4013

Maraganore JM, Bourdon P, Jablonski J, Ramachandran KL, Fenton J (1990) Design and characterization of hirulogs: a novel class of bivalent peptide inhibitors of thrombin. Biochemistry 29:7095–7101

McDowell RS, Gadek TR (1992) Structural studies of potent constrained RGD peptides. J Am Chem Soc 114:9245–9253

McDowell RS, Gadek TR, Barker PL, Burdick DJ, Chan KS, Quan CL, Skelton N, Struble M, Thorsett ED, Tischler M, Tom JY, Webb TR, Burnier JP (1994a) From peptide to non-peptide 1. The elucidation of a bioactive conformation of a arginine-glycine-aspartic acid recognition sequence. J Am Chem Soc 116:5069–5076

McDowell RS, Blackburn BK, Gadek TR, McGee LR, Rawson T, Reynolds ME, Robarge KD, Someers TC, Thorsett ED, Tischler M, Webb II RR, Venuti MC (1994b) From peptide to non-peptide 2. The de novo design of potent, non-peptidal inhibitors of platelet aggregation based on benzodiazepinedione scaffold. J Am Chem Soc 116:5077–5083

Mohamadi F, Richards NG, Guida WC, Liskamp R, Lipton M, Caulfield C, Chang G, Hendrickson T, Still WC (1990) An integrated software system for modeling organic and bioorganic molecules using molecular mechanics. J Comp Chem 11:440–467

Moon JB, Howe WJ (1991) Computer design of bioactive molecules: a method for receptor-based de novo ligand design. Proteins Struct Funct Genet 11:314–328

Moos WH, Green GD, Pavia MR (1993) Recent advances in the generation of molecular diversity. Ann Rep Med Chem 28:315–324

Morimoto H, Murai M, Maeda Y, Hagiwara D, Miyake H, Matsuo M, Fujii T (1992) FR113680: a novel tripeptide substance P antagonist with NK1 receptor selectivity. Br J Pharmacol 106:123–126

Morley JS, Tracy HJ, Gregory RA (1965) Structure-function relationships in the active C-terminal tetrapeptide sequence of gastrin. Nature 207:1356–1359

Müller G, Gurrath M, Kessler H, Timpl R (1992) Dynamic forcing, a method for evaluating activity and selectivity profiles of RGD peptides. Angew Chem (Int Ed Engl) 31:326–328

Nichols AJ, Ruffolo RR, Huffman WF, Poste G, Samanen J (1992) Development of GPIIb/IIIa antagonists as antithrombotic drugs. Trends Pharmacol Sci 13:413–417

Nutt RF, Brady SF, Colton CD, Sisko JT, Ciccarone TM, Levy MR, Duggan ME, Imagire IS, Gould RJ, Anderson PS, Veber DF (1992) Development of novel highly selective fibrinogen receptor antagonists as potentially useful antithrombotic agents. In: Smith JA, Rivier JE (eds) Peptides, chemistry and biology. ESCOM, Leiden, pp 914–916

O'Donnell M, Garippa RJ, O'Neill NC, Bolin DR, Cottrell JM (1991) Structure-activity studies of vasointestinal polypeptide. J Biol Chem 266:6389–6392

O'Neil KT, Hoess RH, Jackson SA, Ramachandran NS, Mousa SA, DeGrado WF (1992) Identification of novel peptide antagonists for GPIIb/IIIa from a conformationally constrained phage peptide library. Proteins Struct Funct Genet 14:509–515

Paladino J, Kucharczyk N, Morris AD, Thibault-Naze M, Mahieu JP, Serkiz B, Volland JP, Autissier C, Fauchère JL (1994) Estimation of blood levels of endothelin and neurokinin receptor antagonists at the rat portal and jugular veins after oral administration as a tool in peptide drug design. Drug Des Discov 12:121–128

Perren TJ, Clayton RN, Blackledge G, Bailey LC, Holder G, Lynch SS, Arkell DG, Cottam J, Farrar D, Young CH (1986) Pharmacokinetic and endocrinological parameters of a slow-release depot preparation of the GnRH analogue ICI 118630 (Zoladex) compared with a subcutaneous bolus and continuous subcutaneous infusion of the same drug in patients with prostatic cancer. Cancer Chemother Pharmacol 18:39–43

Pierschbacher MD, Ruoslahti E (1984) Cell attachment activity of fibronectin can be duplicated by small synthetic fragments of the molecule. Nature 309:30–33

Pless J, Bauer W, Briner U, Doepfner W, Marbach P, Maurer R, Petcher TJ, Reubi JC, Vonderscher J (1986) Chemistry and pharmacology of SMS

201995, a long-acting octapeptide analogue of somatostatin. Scand J Gastroenterol 21 [S119]:54–64

Pliska V (1981) Semiempirical structrue-activity relationships in peptide pharmacology. In: Eberle, Geiger, Wieland (eds) Perspectives in peptide chemistry, Karger, Basel, pp 221–236

Polarek JW, Clark RA, Pickett MP, Pierschbacher MD (1994) Development of a provisional extracellular matrix to promote wound healing. Wounds 6:46–53

Rao SN (1992) Bioactive conformation of Arg-Gly-Asp by X-ray data analyses and molecular mechanics. Peptide Res 5:148–155

Reed J, Hull WE, Lieth CW, K.bler D, Suhai S, Kinzel V (1988) Secondary structure of the Arg-Gly-Asp recognition site in proteins involved in cell-surface adhesion. Eur J Biochem 178:141–154

Regoli D, Boudon A, Fauchère JL (1994) Receptors and antagonists for substance P and related peptides. Pharmacol Rev 46:551–599

Reich SH (1994) Protein crystal structure-based design of novel non peptide HIV protease inhibitors. Presented at the meeting New advances in peptidomimetics and small molecule design for drug development, Philadelphia, 23–25 March, abstract 330. AGOURON, San Diego

Reich SH, Webber SE (1993) Structure-based drug design (SBDD): every structure tells a story. Perspect Drug Discov Design 1:371–390

Richardson DW, Robinson AG (1985) Desmopressin. Ann Intern Med 103:228–239

Rittel W (1973) Synacthen depot. In: Schuppli R (ed) ACTH, eine Standortsbestimmung für die Praxis. Huber, Bern, pp 11–18

Rose GD (1978) Prediction of chain turns in globular proteins on a hydrophobic basis. Nature 272:586–590

Rosenberg SH, Spina KP, Woods KW, Polakowski J, Martin DL, Yao Z, Stein HH, Cohen J, Barlow JR, Egan DA, Tricarico KA, Baker WR, Kleinert HD (1993) Studies directed toward the design of orally active renin inhibitors. Some factors influencing the absorption of small peptides. J Med Chem 36:449–459

Samanen J, Ali F, Romoff T, Calvo R, Sorenson E, Vasko J, Storer B, Berry D, Bennett D, Strohsacker M, Powers D, Stadel J, Nichols A (1991) Development of a small RGD peptide fibrinogen receptor antagonist with potent antiaggregatory activity in vitro. J Med Chem 34:3114–3125

Sanderson PM, Glen RC, Payne AW, Hudson BD, Heide C, Tranter GE, Doyle PM, Harris CJ (1994) Characterization of the solution conformation of a cyclic RGD peptide analogue by NMR spectroscopy allied with a genetic algorithm approach and constrained molecular dynamics. Int J Peptide Protein Res 43:588–596

Scarborough RM, Naughton MA, Teng W, Rose JW, Phillips DR, Nannizzi L, Arfsten A, Campbell AM, Charo IF (1993) Design of potent and specific integrin antagonists. J Biol Chem 268:1066–1073

Scott JK, Smith GP (1990) Searching for peptide ligands with an epitope library. Science 249:386–390

Schiffer M, Edmundson AB (1967) Use of helical wheels to represent the structures of proteins and to identify segments with helical potential. Biophys J 7:121–135

Schütz W, Freissmuth M (1992) Reverse intrinsic activity of antagonists on G protein coupled receptors. Trends Pharmacol Sci 13:376–380

Schwyzer R (1973) Molecular mechanisms of polypeptide hormone action. In: Hanson H, Jakubke HD (eds) Peptides 1972. North Holland/Elsevier, New York, pp 424–436

Siahaan T, Lark LR, Pierschbacher M, Ruoslahti E, Gierasch LM (1990) A conformationally constrained RGD analog specific for the vitronectin receptor: a model for receptor binding. In: Rivier JE, Marshall GR (eds) Peptides, chemistry, structure and biology. ESCOM, Leiden, pp 699–701

Sibanda BL, Blundell T, Hobart PM, Fogliano M, Bindra JS, Dominy BW, Chirgwin JM (1984) Computer graphics modelling of human renin. Specificity, catalytic activity and intron-exon junctions. FEBS Lett 174:102–111

Simoons ML, de Boer J, van den Brand MJ, van Miltenburg AJ, Hoorntje JC, Heyndrickx GR, van der Wieken LR, De Buono D, Rutsch W, Schaible TF, Weisman HF, Klootwijk P, Nijssen KM, Stibbe J, de Feyter PJ (1994) Randomized trial of a GPIIb/IIIa platelet receptor blocker in refractory unstable angina. Circulation 89:596–603

Snider RM, Constantine JW, Lowe JA III, Longo KP, Lebel WS, Woody HA, Drozda SE, Desai MC, Vinick FJ, Spencer RW, Hess HJ (1991) A potent non peptide antagonist of substance P (NK-1) receptors. Science 251:435–437

Thurieau C, Félétou M, Canet E, Fauchère JL (1994a) p-Guanidino-benzoyl-[Hyp3,Thi5, D-Tic7,Oic8]bradykinin is almost completely devoid of the agonist effect of HOE140 on the endothelium-free femoral artery of sheep. Bioorg Med Chem Lett 4:781–784

Thurieau C, Guyard C, Simonet S, Verbeuren TJ, Fauchère JL (1994b) Synthesis of a new bivalent hirudin analog (Hirufos) which includes a stable 4'-phosphono-L-phenylalanine mimic of (L-tyrosine O4-sulfate)-63. Helv Chim Acta 77:679–684

Toniolo C, Benedetti E (1991) The polypeptide 3_{10}-helix. Trends Biochem Sci 16:350–353

Trumpp-Kallmeyer S, Hoflack J, Hibert M (1994) Modeling of G-protein coupled receptors: application to the NK1 receptor. In: Buck SH (ed) The tachykinin receptors. Humana, Totowa, pp 237–255

Veber DF (1992) Design and discovery in the development of peptide analogs. In: Smith JA, Rivier JE (eds) Peptides, chemistry and biology, ESCOM, Leiden, pp 3–14

Verlander MS, Goud AN, Makineni R (1991) Problems and projected costs for the preparation of bulk quantities of peptides for human administration. In: Bloom SR, Burnstock G (eds) Peptides: a target for new drug development. IBC Technical Services, London, pp 135–146

Wallace BA, Ravikumar K (1988) The gramicidin pore: crystal structure of a cesium complex. Science 241:182–187

Watling KJ (1992) Nonpeptide antagonists herald new era in tachykinin research. Trends Pharmacol Sci 13:266–269

Weber C, Wider G, von Freyberg B, Traber R, Braun W, Widmer H, Wüthrich K (1991) The NMR structure of cyclosporin A bound to cyclophilin in aqueous solution. Biochemistry 30:6563–6574

Weiner PK, Langridge R, Blaney JM, Schaefer R, Kollman PA (1982) Electrostatic potential surfaces. Proc Natl Acad Sci USA 79: 3754–3758

Wenger R (1984) Synthesis of cyclosporins. Helv Chim Acta 67:502–525

Wirth K, Hock FG, Albus U, Linz W, Alperman HG, Anagnostopoulos H, Henke S, Breipohl G, König W, Knolle J, Schölkens BA (1991) HOE140 a new potent and long-acting bradykinin antagonist: in vivo studies. Br J Pharmacol 102:774–777

Wlodawer A, Erickson JW (1993) Structure-based inhibitors of HIV-1 protease. Annu Rev Biochem 62:543–585

Zablocki JA, Miyano M, Garland RB, Pireh D, Schretzman L, Rao SN, Lindmark RJ, Panzer SG, Nicholson NS, Taite BB, Salyers AK, King LW, Campion JG, Feigen LP (1993) Potent in vitro and in vivo inhibitors of platelet aggregation based upon the Arg-Gly-Asp-Phe sequence of fibrinogen. J Med Chem 36:1811–1819

Zablocki JA, Bovy PR, Rico JG, Rogers TE, Lindmark RJ, Tjoeng FS (1994) Potent in vitro and in vivo inhibitors of platelet aggregation based upon the Arg-Gly-Asp-Phe sequence of fibrinogen. A lead series of orally active antiplatelet agents. Abstract paper of the American Chemical Society, 207th meeting, part 1, MEDI 14, Coden ACSRAL ISSN:0065–7727

Zuckermann RN, Martin EJ, Spellmeyer DC, Stauber GB, Shoemaker KR, Kerr JM, Figliozzi GM, Goff DA, Siani MA, Simon RJ, Banville SC, Brown EG, Wang L, Richter LS, Moos WH (1994) Discovery of nanomolar ligands for 7-transmembrane G-protein-coupled receptors from a diverse N-substituted-glycine peptoid library. J Med Chem 37:2678-2685

9 Quantitative Structure–Activity Relationships and Crystallography in Industrial Drug Design

H. Kubinyi

9.1 Introduction

Industrial drug research differs in many respects from research at universities and research institutes (e.g., Gund et al. 1992). The primary goal of the chemists, biochemists, biologists, and theoreticians involved in a therapeutic project is the development of new and better drugs with the least effort, especially of human resources. Thus, quantitative structure–activity relationships (QSAR) and modeling groups in industry are well equipped with the most efficient hardware and software. But their research is not directed to their own projects, it accompanies the ongoing activities of the different drug development teams. To avoid misunderstandings: a lot of methodological progress in QSAR and modeling comes from industrial groups, but this is more a side effect than the main goal of industrial research. The reasons for the development of new methods are manifold. Sometimes commercial software produces results which are not reliable and cannot be accepted if objective criteria are applied; some other methods are developed because commercial tools are not yet available.

Predominant features of an active drug are a certain range and pattern of lipophilicity which enable the drug to arrive at and to interact with its binding site and a complementarity of its three-dimensional (3D) structure to this binding site, in geometry as well as in physicochemical properties (e.g., Kubinyi 1993a,b).

Mass screening of thousands or tens of thousands of compounds in molecular test systems, using human proteins and receptors, offers rapid access to new lead structures. In addition, structure-based and computer-aided design are well-established techniques in the search for new leads and their rational optimization. However, affinity to the binding site is a necessary but not a sufficient property of a drug. Bioavailability, metabolic stability, a certain duration of action, selectivity, an acceptable therapeutic ratio, and the lack of intolerable side effects are the main criteria for its therapeutic utility.

Some of these properties, e.g., rare side effects, cannot be systematically investigated nor can they be described by mathematical models. Thus, they cannot be predicted. Others, such as bioavailability and tissue distribution, mainly depend on lipophilicity if no carriers mediate drug absorption and transport in the biological system.

9.2 Lipophilicity and Drug Design

The predominant role of lipophilicity for drug transport is well understood (e.g., Kubinyi 1979a,b, 1993a; Dearden 1990). Many linear and nonlinear relationships between lipophilicity and biological activities have been described (Figs. 1,2).

Nonlinear lipophilicity–activity relationships can be explained by the kinetics of drug transport as well as by the assumption of an equilibrium between phases of different lipophilicity (Kubinyi 1979a,b).

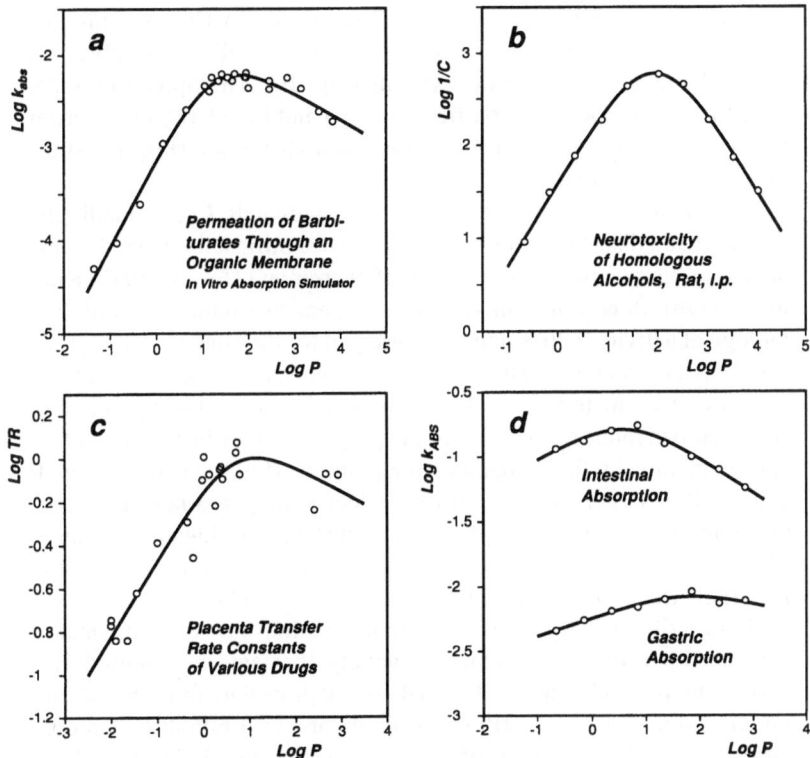

Fig. 1a–d. Nonlinear lipophilicity relationships. **a** Permeation of barbiturates in an vitro absorption simulator (Kubinyi 1979a); **b** neurotoxicity of homologous primary alcohols in the rat, i.p. application (Kubinyi 1979a); **c** placenta transfer rate constants of various drugs (Akbaraly et al. 1985); **d** gastric and intestinal absorption of carbamates in the rat (Kubinyi 1979a)

Even the rate constants of substance transport in simple two- or three-compartment n-octanol/water systems show such nonlinear dependences. For polar compounds, the transport rate constants from an aqueous phase into an organic phase increase with increasing lipophilicity. However, this increase of rate constants is limited by diffusion. For lipophilic compounds, the rate constants approach a constant value. The reverse situation holds for the transport rate constants from the organic phase into the aqueous phase (Kubinyi 1978). Correspondingly, nonlinear relationships of drug transport on lipophilicity are observed even in simple in vitro model systems (Fig. 1a). The lipophilicity dependences of blood–brain barrier (Fig. 1b) and blood–placenta barrier transport (Fig. 1c) and gastric and intestinal absorption (Fig. 1d) show the same characteristics.

Ions are much more polar than neutral compounds. Correspondingly, transport and distribution of acids and bases significantly depend on the degree of dissociation and ionization (Scherrer and Howard 1977; Kubinyi 1979b). In equilibrium systems (e.g., enzyme inhibition, Fig. 2a), biological activity values must be corrected for the concentration of the active species. In kinetically controlled systems, the lipophilicity values have to be corrected. If acids and basis of comparable lipophilicity cover a broad range of pK_a values, the pK_a values may be taken as a first approximation of their apparent lipophilicity (Fig. 2b). If the log P values differ largely, the use of pK_a values is inappropriate. Transport rate constants at different pH values must be considered separately (Fig. 2c) or ionization corrected, apparent partition coefficients, log P_{app} (Scherrer and Howard 1977), have to be used (Fig. 2d).

To describe such lipophilicity–activity relationships, experimental or calculated partition coefficients and acidity constants are required. Experimental methods, such as the shake-flask procedure or high pressure liquid chromatography (HPLC) methods are well-established procedures for the determination of partition coefficients. Several methods were described for the simultaneous determination of pK_a values and n-octanol/water partition coefficients (e.g., Avdeef 1992, 1993).

For large series of compounds, the partition coefficients of simple organic compounds can be calculated by the program CLOGP (Leo 1990), by other hydrophobic fragmental constant methods (Rekker 1977; Rekker and Mannhold 1992), or by atom-based approaches (Ghose and Crippen 1986, 1987; Suzuki and Kudo 1990; Suzuki 1991; Convard et al. 1994). Polar compounds and compounds which bear electronegative groups that are not parameterized in the programs still offer problems.

Comparable restrictions apply to the calculation of the pK_a values of acids and bases. With the exception of simple analogs of aliphatic and aromatic acids, phenols or amines, the quantitative estimation of pK_a values is still a difficult and basically unsolved problem.

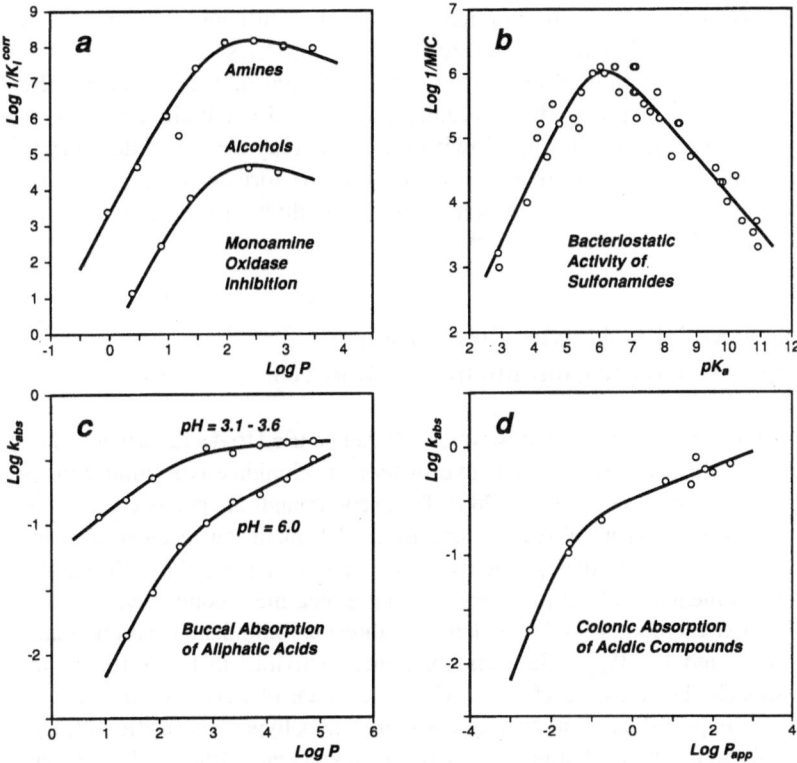

Fig. 2a–d. Nonlinear lipophilicity relationships. **a** Monoamine oxidase inhibition of amines and alcohols (Kubinyi 1979b); **b** bacteriostatic activity of acidic and basic sulfonamides (Silipo and Vittoria 1979); **c** buccal absorption of homologous aliphatic acids at different pH values (Kubinyi 1979a); **d** colonic absorption of acidic compounds (Kubinyi 1979a)

Most often lipophilicity optima are significantly different in simple in vitro systems (also in tissue cultures and isolated organs) and in vivo; this important fact has to be taken into account even in early phases of drug optimization. Frustrations arise from the violation of this principle, if "rational" drug design is guided only by in vitro experiments. Often toxicity and activity show different lipophilicity optima; if this observation is to be used for the design of safer drugs, it is important to check whether it is a real difference or whether it is only caused by different time scales, different biological test models, or different metabolism in different species.

Many lipophilic drugs have CNS side effects and show nonspecific toxicity. Lipophilic drugs readily pass the blood–brain and blood–placenta barriers, they are slowly eliminated, they are more inhibitory to many enzymes, and reactive species may be formed in their metabolism; thus, as a general principle, drugs should be made as hydrophilic as possible (Hansch et al. 1987).

9.3 Activity–Activity Relationships: QSAR and the Continuation of Research

An interesting example of how QSAR can be used to decide when to stop research in a project is an investigation of clonidine-type imidazolines (Timmermans et al. 1981, 1984). These compounds are potent α_2 receptor agonists, lowering blood pressure by a CNS-mediated mechanism. Clonidine itself is in therapeutic use as an antihypertensive drug. Beside the CNS-mediated blood pressure-lowering effect, these compounds cause a peripheral blood vessel contraction by direct α_2-stimulation. Timmermans measured log P_{app} values and biological activities in two different rat models. Hypotensive effects (pC_{25} values) are observed in anesthetized rats. In pithed rats, without any CNS regulation, hypertensive effects (pC_{60} values) result. In addition, in vitro α_1 and α_2 receptor affinities, $IC_{50}\alpha_1$ and $IC_{50}\alpha_2$, were measured. Three important conclusions can be drawn from the resulting equations (Eqs. 1–3) (Kubinyi 1988).

$$pC_{60} = 1.163 \ (\pm 0.21) \log 1/IC_{50}\alpha_2 - 0.962 \ (\pm 0.39) \tag{1}$$
$$(n = 21; r = 0.936; s = 0.317; F = 135.15)$$

$$pC_{25} = 0.805 \ (\pm 0.22) \log P - 3.373 \ (\pm 1.02) \log(\beta P + 1)$$
$$+ 1.071 \ (\pm 0.20) \log 1/IC_{50}\alpha_2 - 1.164 \ (\pm 0.39) \tag{2}$$
$$\log \beta = -1.986 \qquad\qquad \text{optimum log P} = 1.48$$
$$(n = 21; r = 0.971; s = 0.284; F = 65.22)$$

$$pC_{25} = 0.784 \ (\pm 0.26) \log P - 3.685 \ (\pm 1.39) \log(\beta P + 1)$$
$$+ 0.830 \ (\pm 0.20) \ pC_{60} - 0.189 \ (+0.30) \tag{3}$$
$$\log \beta = -2.078 \qquad\qquad \text{optimum log P} = 1.51$$
$$(n = 21; r = 0.954; s = 0.354; F = 40.52)$$

First, the undesired hypertensive side effect (pC_{60}) is a linear function of α_2-affinities (Eq. 1). Second, the CNS-mediated hypotensive effect (pC_{25}) is also a linear function of α_2 affinities but a nonlinear lipophilicity relationship has to be included (Eq. 2). The drugs must pass the blood–brain barrier on their way from the site of application to the site of action. Hypertensive and hypotensive activities can only partially be separated (Eq. 3). Analogs within the lipophilicity optimum will be antihypertensives because the CNS-mediated effect predominates, whereas for polar analogs the peripheral effect will be observed. Using Eqs. 1 and 2, the in vivo activities of further analogs can be predicted from the results in a simple in vitro system. However, in this class of compounds no improvement can be expected from further structural modification; research activities could be stopped.

Other examples of the role of QSAR in deciding on the continuation of research were given by Corwin Hansch. In a series of antitumor nitrosoureas, different lipophilicity optima were observed for antitumor activity and for toxicity, which implies that further research should be performed on less lipophilic and therefore less toxic analogs (Hansch et al. 1980). For antitumor aryl triazenes, a nonlinear lipophilicity–activity relationship and a significant influence of electronic properties gave a clear indication to continue research on analogs with a lipophilicity range around log P = 1.0–1.5 and electron donor substituents in the aromatic ring. However, as the p-OMe analog in aqueous solution already has a chemical half-life time of only about 12 min, the QSAR result was the signal to stop research in this area (Hatheway et al. 1978; Hansch et al. 1978).

9.4 Crystallography and Biologically Active Conformations of Drugs

For calcium-antagonistic verapamil (formula 1) analogs, a significant nonlinear dependence of calcium channel binding on lipophilicity is observed. The only exceptions are two analogs, where the isopropyl group of verapamil is replaced by a hydrogen atom (Fig. 3) (Kubinyi and Klebe 1987; König 1987).

In searching for the reasons of this unexpected result, we first looked at the X-ray structure of verapamil. Verapamil is a molecule with many

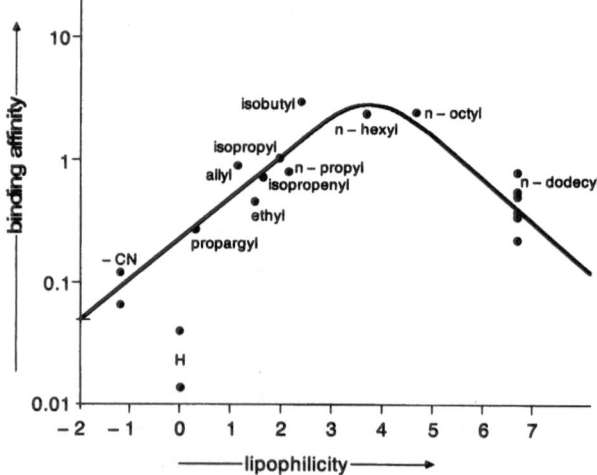

(1)

internal rotational degrees of freedom, thus its conformation at the binding site may be very different from the conformation in the crystal. The same restriction applies to 2D nuclear magnetic resonance (NMR) structures in solution. The problem is: at the receptor site a flexible fit occurs, induced by the interactions with the binding site. The resulting drug conformation is the one which allows the largest number of stabilizing interactions, irrespective of whether this induced-fit conformation is indeed the lowest-energy conformation or not. On the other hand, one conformational feature is of special importance for the binding of verapamil: the nitrile group and the aromatic ring are arranged in a plane, induced by the bulky isopropyl group and by the chain bearing the rest of the molecule. The rotational barrier seems to be high enough to stabilize this conformation at the receptor site, too. Even for an

Fig. 3. Nonlinear relationship between calcium channel binding affinities of verapamil analogs and lipophilicity. (Kubinyi and Klebe 1987)

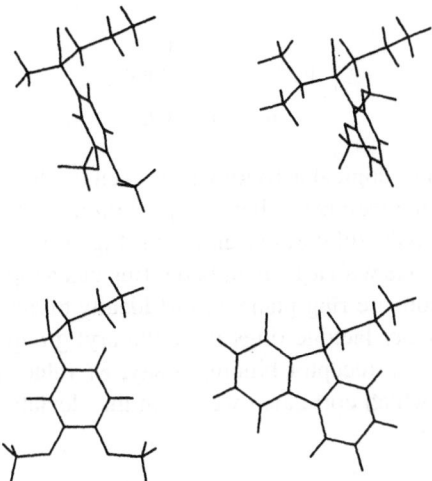

Fig. 4. Stereochemistry of verapamil (formula 1) analogs (only the left parts of the analogs are shown). *Upper left*: methyl (instead of isopropyl) analog. *Upper right*: 2,6-dimethoxy (instead of 3,4-dimethoxy) analog. *Lower left*: hydrogen (instead of isopropyl) analog. *Lower right*: fluorene analog. (Höltje and Hense 1985; Gualtieri et al. 1985; Kubinyi and Klebe 1987)

analog with methyl instead of isopropyl (Kubinyi and Klebe 1987) and for a 2,6-dimethoxy analog of verapamil (Fig. 4) (Höltje and Hense 1985) this coplanar conformation is the most probable one. On the other hand, for the analog with hydrogen instead of isopropyl such a conformation is no longer the most stable one. In the lowest energy conformation, the plane of the aromatic ring intersects the angle between the C-H bond and C-C≡N bond, explaining the much lower than predicted affinity of this analog. Evidence for this explanation comes from a fluorene analog (Fig. 4), where the nitrile group and the benzene rings are in different planes and which lacks calcium-antagonistic activity (Gualtieri et al. 1985).

Another calcium antagonist, nifedipine (formula 2), has a much more rigid structure. Due to the bulky ester groups at both sides of the dihydropyridine ring and due to the o-NO_2 group, the plane of the aromatic ring and the plane of the dihydropyridine ring form an angle of about 90°. That this orthogonal arrangement of both rings is essential

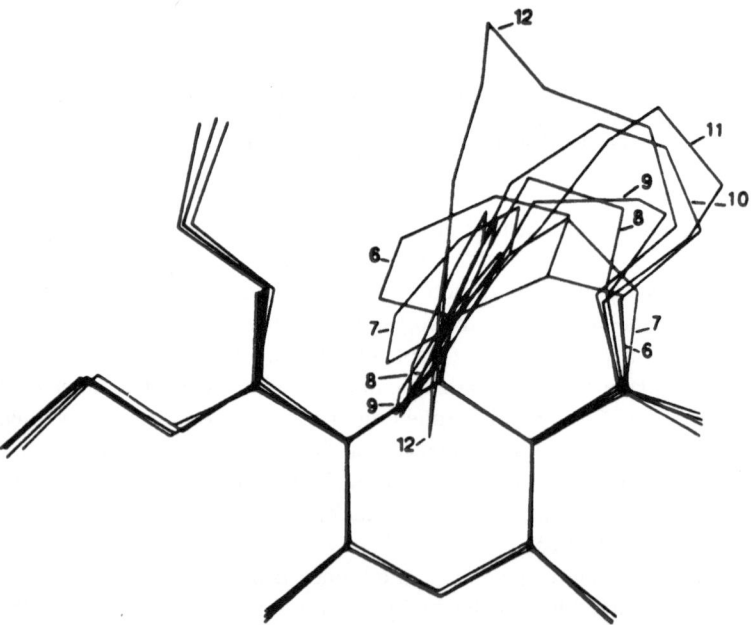

$$H_3COOC \qquad COOCH_3 \qquad (2)$$

$$H_3C \qquad \underset{H}{N} \qquad CH_3$$

for binding and biological activity was proven by the synthesis of 6- to 12-membered ring lactones, where the phenyl ring is frozen at different angles relative to the dihydropyridine ring (Fig. 5) (Seidel et al. 1985).

Only the lactone with a 12-membered ring can adopt a conformation in which the aromatic ring plane almost ideally bisects the dihydropyridine ring. Smaller lactone rings force the aryl group to deviate from this geometry. In a receptor binding assay, K_i values for the lactones were observed which correlated well with this deviation, expressed by $\Delta\alpha$ (Eq. 4; Fig. 6)

Fig. 5. Superimposed X-ray structures of dihydropyridine lactones. *Numbers* indicate the different ring sizes. (Seidel et al. 1985)

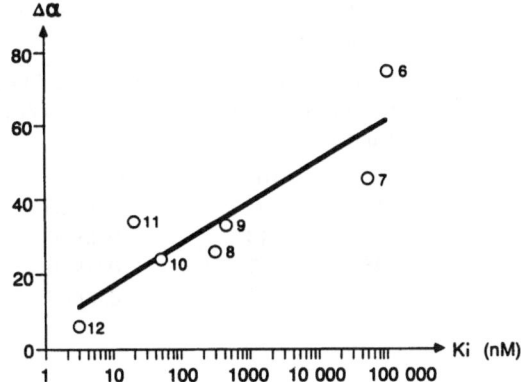

Fig. 6. Structure–activity relationships of dihydropyridine lactones. *Numbers* indicate the different ring sizes. (Seidel et al. 1985)

$$\log K_i = 0.067 \ (\pm 0.017) \ \Delta\alpha + 0.19 \ (\pm 0.34) \qquad (4)$$
$$(n = 7; \ r = 0.88; \ F = 16.5)$$

Alternatively, biologically active conformations can be derived by the active analog approach (Marshall and Naylor 1990; Marshall 1993) and, of course, from X-ray studies of ligand–protein complexes and dedicated 2D NMR investigations.

9.5 Drug–Receptor Interactions

About 20 years ago, Höltje and Kier started to calculate interaction energies between hypothetical receptor models and various ligands (Höltje and Kier 1974). From the comparison of the interaction energies of different acetylcholine analogs (Table 1) with different amino acid side chains they concluded that the cationic head of acetylcholine and its charged and neutral analogs bind to an aromatic ring (Höltje and Kier 1975), not to an anionic site of cholinesterase, as postulated before. Many years later, this hypothesis was confirmed by the 3D structure determination of acetylcholinesterase, which indeed shows a hydrophobic aromatic environment (Sussman et al. 1991).

Table 1. Interaction energies between actylcholine and its analogs and different amino acid side chain mimics. (Höltje and Kier 1975)

R-CH_2-CH_2-$OCOCH_3$	Cholinesterase, exp. hydrolysis rates (%)		Theoretical model, interaction energies	
	Human	Horse	Acetate	Benzene
N+$(CH_3)_3$	100	100	100	100
C$(CH_3)_3$	60	24	0.00004	25
CH$(CH_3)_2$	24	14	0.00002	18
CH_2CH_3	16	7	0.00001	13
CH_3	10	3	0.00001	10

Also for verapamil analogs a receptor model could be derived which indicated an interaction between the substituents of the left aromatic ring (the benzyl nitrile part of 1) with the charged guanidinium side chain of an arginine (Eq. 5; IE = calculated interaction energies in standard geometries) (Höltje 1982).

$$\log 1/ED_{50} = 1.14 \, IE - 0.97 \tag{5}$$
$$(n = 10; r = 0.973; s = 0.095)$$

Hypothetical interaction models, including one or even several amino acid side chains are, of course, rough simplifications of the real environment of a ligand. But nature tells us that one amino acid can indeed make the difference between rats and humans. Although the 5-HT_{1B} receptor proteins of both species have about 90% amino acid identity, there is no significant relationship in the binding affinities of several serotonin and β receptor ligands ($r = 0.27$). If only one amino acid of the human wild type protein, 355-Thr, is changed to Asn, the corresponding amino acid of the rat receptor, this human receptor mutant now behaves like a rat receptor ($r = 0.98$) (Parker et al. 1993).

The importance of human proteins in biological test systems can also be illustrated by recent results on renin inhibition (Clozel and Fischli 1993). The K_i values of remikiren, for human renin and for renins from two different monkey species, are in the range of about 1–2 nM. In contrast, the renins of dogs and rats, which both are typical test animals in cardiovascular research, are much less sensitive; the K_i values are 107 nM and 3600 nM, respectively.

9.6 QSAR and Protein Crystallography

In some cases the results of QSAR studies can be related to the 3D structures of the ligand binding sites. This has been done, for example, for dihydrofolate reductase (DHFR), papain and other cysteine proteases, trypsin and other serine proteases, and acetylcholinesterase (Blaney and Hansch 1990; Selassie and Klein 1993).

For the inhibition of *Escherichia coli* DHFR (Eq. 6) and *Lactobacillus casei* DHFR (Eq. 7) by benzylpyrimidines, two closely related QSAR models were derived (Eqs. 6,7) (Hansch et al. 1982).

$$\log 1/K_{i\ app} = 0.75\ (\pm 0.26)\ \pi_{3,4,5} - 1.07\ (\pm 0.34)\ \log\ (\beta \cdot 10^{\pi_{3,4,5}} + 1)$$
$$+ 1.36\ (\pm 0.24)\ MR_{3,5} + 0.88\ (\pm 0.29)\ MR_4$$
$$+ 6.20 \tag{6}$$
$$\log \beta = 0.12 \qquad\qquad \text{optimum } \pi = 0.25$$
$$(n = 43;\ r = 0.903;\ s = 0.290)$$

$$\log 1/K_{i\ app} = 0.31\ (\pm 0.11)\ \pi_{3,4} - 0.88\ (\pm 0.24)\ \log\ (\beta \cdot 10^{\pi_{3,4}} + 1)$$
$$+ 0.95\ (\pm 0.21)\ MR_{3,4} + 5.32 \tag{7}$$
$$\log \beta = -1.33 \qquad\qquad \text{optimum } \pi = 1.05$$

The only difference between both equations is that the 5-substituents of the benzyl group contribute to biological activities in the case of *E. coli* DHFR, while they have no effect in the case of *L. casei* DHFR. An explanation could be given as soon as the X-ray structures of the enzymes were available. Both have about the same geometry of the binding site but a rigid leucine side chain in the *L. casei* DHFR forms a much narrower cleft than the more flexible methione side chain in *E. coli* DHFR (Hansch et al. 1982).

Paul Bartlett investigated the inhibition of thermolysin by different analogs of the inhibitors (formula 3) (X = –NH–, R = OH, Gly-OH, Ala-OH, Leu-OH, Phe-OH) (Bartlett and Marlowe 1987).

$$(3)$$

While the phosphonamidates are potent inhibitors, the isosteric phosphonates $(X = -O-)$ are much less active. On the other hand, phosphinates $(X = -CH_2-)$ are about as active as the phosphonamidates. An explanation could be derived from the known X-ray structure of the thermolysin/phosphonamidate complex. Only the phosphonamidates form a hydrogen bond between the –NH– group and an alanine backbone C=O group. In addition to the lack of the hydrogen bond, there is an unfavorable O···O interaction in the phosphonates and an unfavorable effect of desolvation of the –X– group in the formation of the complex. This latter unfavorable effect applies to the phosphonates and the phosphonamidates. Although the –CH$_2$– group of the phosphinates cannot form a hydrogen bond to the protein, there are no unfavorable effects of desolvation and O···O repulsion, explaining their high affinities (Morgan et al. 1991).

In the binding of Cbz-Phe and its –CH$_2$– analog, β-phenylpropionyl-Phe, to thermolysin, the system compensates by a completely different binding mode of both analogs (Kester and Matthews 1977).

QSAR model building and rational drug design may become extremely difficult or even impossible if the 3D structure of the binding site is unknown, which is a common situation in the application of QSAR or 3D QSAR methods. A purine nucleoside phosphorylase inhibitor showed increased inhibitory activity when N-9 was replaced by C and when an 8-NH$_2$ group was introduced into the system. The logical consequence, to combine both structural features, does not produce the expected increase of inhibitory activity, a relative decrease of activities is observed instead. An explanation could be derived from the X-ray structures of the inhibitor complexes (Fig. 7). The change of N-7 from a hydrogen bond donor to an acceptor group leads to a much more favorable geometry of the ligand–protein interaction. However, this shift causes the 242-Thr oxygen atom to move away from the 8-NH$_2$ group; an unfavorable polar–nonpolar interaction and a partial van der Waals overlap with the methyl group of the threonine results (Montgomery et al. 1993; Montgomery and Niwas 1993).

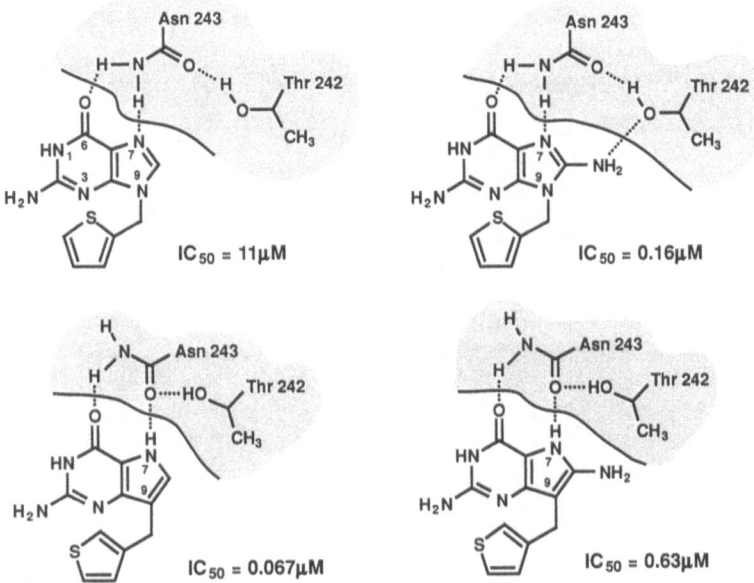

Fig. 7. Binding geometries and structure–activity relationships of purine nucleoside phosphorylase inhibitors. (Montgomery et al. 1993; Montgomery and Niwas 1993)

9.7 Structure-Based and Computer-Assisted Design

Structure-based design of ligands is now a well-established technique which is applied by many pharmaceutical companies (e.g., Beddell 1992; Navia and Murcko 1992; Borman 1992; Reich and Webber 1993). Three-dimensional searching techniques and de novo design are newer approaches, which significantly reduce the effort in syntheses and testing of new analogs.

Among several competing approaches, the de novo design program LUDI (Fig. 8), developed by Hans-Joachim Böhm at BASF (Böhm 1992a,b, 1993), has received a broad acceptance in pharmaceutical industry because of its user-friendliness and its several different features.

LUDI starts from the 3D structure of a protein and identifies functional groups and hydrophobic areas within a predefined binding site.

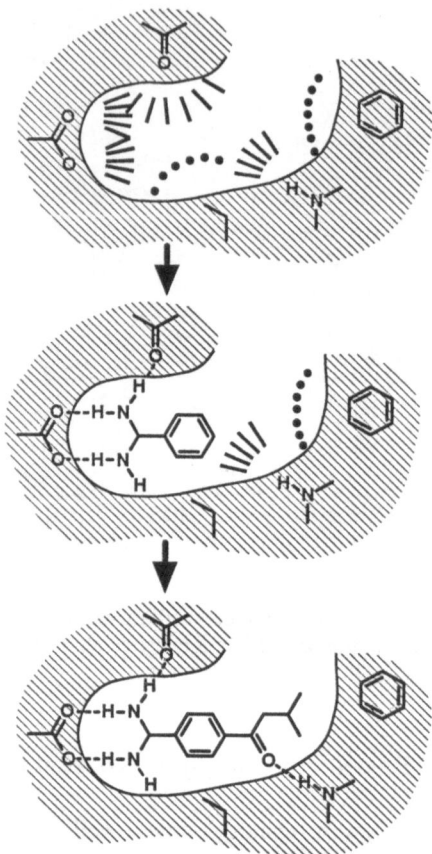

Fig. 8. Concept of the LUDI program for the de novo design of enzyme inhibitors (Böhm 1992a,b, 1993). The program automatically assigns the optimal positions of interacting groups (*top*), searches for ligands and sorts the hits (*middle*, a good hit), and attaches residues to these hits or to other lead structures (*bottom*)

The program assigns positions of hydrogen-bond donor and acceptor groups, as well as liphophilic aliphatic and aromatic groups, which are capable of interacting with the functional and hydrophobic groups of the protein. Then an internal or any external 3D structural database is searched for ligands which fit the binding site. Hits are sorted by a scoring function (Eq. 8) which was derived from the 3D structure-based interpretation of the binding constants of several ligand–protein complexes (Böhm 1994).

$$\log 1/K_i = 1.4\,(\pm\,0.4)\text{ Ionic} + 0.83\,(\pm\,0.3)\text{ HBonds}$$
$$+ 0.030\,(\pm\,0.01)\text{ Lipo} - 0.25\,(\pm\,0.1)\text{ nRot} - 0.91\,(\pm\,1.4) \tag{8}$$
$$(n = 45;\ r = 0.875;\ s = 1.395;\ F = 32.80)$$

Additional features of LUDI are the attachment of functional groups to already existing lead structures and an active analog mode, where only the superimposed 3D structures of active ligands are used as a template for the design of other potential ligands.

9.8 Industrial Drug Design:
The Search for Thrombin Inhibitors

The use of LUDI is illustrated by our work on the design of thrombin inhibitors. Thrombin plays a central role in blood coagulation by mediating the cleavage of fibrinogen to fibrin, which after further processing, together with blood platelets and erythrocytes, forms an insoluble clot. This physiological process is desired in wound healing but life-threatening in stroke, cardiac insult, and other diseases with increased blood coagulation tendency.

Our research in this field started from hirudin, the anticoagulant principle of the leech, *Hirudo medicinalis*. Hirudin is a 65 amino acid protein with no enteral bioavailability and a short duration of action. Thus, research activities were concentrated in three directions: first, to design a hirudin analog with an increased biological half-life; second, to design a short peptide with about the same activity as hirudin itself; third, to design highly selective, small molecule active site inhibitors with sufficient bioavailability, acceptable metabolic stability, and no serious side effects.

Table 2. Biological activities and plasma half-life times of recombinant hirudin, its C-terminal decapeptide (Phe-Glu-Glu-Ile-Pro-Glu-Glu-Tyr-Leu-Gln-OH), the N-succinoylated decapeptide, and LU 58463 (Suc-Tyr-Glu-Pro-Ile-Hyp-Glu-Glu-Smp-Cha-Gln-OH) (Bernard et al. 1994)

Compound	TT $(EC_{100})^a$	AVS $(ED_{15})^b$	$t_{1/2}$ $(min)^c$
r-Hirudin	22	1.42	35
Decapeptide	46 000	ND	ND
N-Suc-decapeptide	5640	ND	ND
LU 58463	80	0.49	57

r-Hirudin, recombinant hirudin; Hyp, hydroxyproline; Smp, *p*-sulfomethyl-phenylalanine; Cha, cyclohexylalanine; ND, not determined.

[a] Prolongation of clotting time (thrombin time, TT), EC_{100} in nmol/l.

[b] Prolongation of arteriovenous shunt patency (AVS), ED_{15} in mg/kg, rat.

[c] Plasma concentration half-life time, in minutes.

Rational design started from the 3D structure of the hirudin–thrombin complex, which we received from the group of Robert Huber and Wolfram Bode at the Max Planck-Institute of Biochemistry in München-Martinsried (Rydel et al. 1990, 1991).

Our first goal was easy to realize. After modification of some amino acids which participate in the binding and others which are not involved, we designed and synthesized a polyethylene glycol (PEG) derivative where the PEG side chains do not interfere with binding and which shows a significant increase in its duration of action. This derivative is now under clinical development.

The second goal was also easy to achieve. It was already known that a C-terminal decapeptide binds to an anionic exosite of thrombin and prevents the binding of fibrinogen; however, this decapaptide is about 2000 times less inhibitory than recombinant hirudin (r-hirudin). Based on the 3D structure of the complex, N-acylated analogs of the C-terminal decapeptide were synthesized and tested for anticoagulant activity. An analog with several modified amino acids, LU 58463, showed a 600-fold increase in inhibitory activity, as compared to the C-terminal decapeptide of r-hirudin. Also in other tests, LU 58463 is as active as r-hirudin and has a slightly prolonged plasma half-life time (Table 2) (Bernard et al. 1994).

For our third goal, the development of an orally active thrombin inhibitor, LUDI designed several analogs, new leads as well as derivatives of already known lead structures. Although many analogs are highly selective inhibitors, in a nanomolar activity range, none of the LUDI-designed compounds have so far shown significant bioavailability, demonstrating once more one of the principal difficulties in rational drug design.

9.9 Summary and Conclusions

A ligand is not a drug. An active inhibitor may have insufficient bioavailability, may be toxic or cause problems in chronic toxicity, may be mutagenic or carcinogenic, may have other serious side effects, or even lack efficacy in the clinics, to mention only the most common reasons for failure in preclinical and clinical drug development.

What is the real value of QSAR, modeling, protein crystallography, structure- based and computer-assisted drug design in industrial research? Different answers can be given. Several success stories of rational drug development have been published (for reviews see, e.g., Boyd 1990; Fujita 1990, 1992; Topliss 1993). More important seems to be that, as a long-term evolutionary process, the QSAR paradigm focused the attention of medicinal chemists to the influence of lipophilicity, polarizability, electron donor and acceptor properties of substituents, and 3D structures of drugs on biological activities.

Research should be guided by a synergism of theory, experimental design, intuition, and the creativity of organic and theoretic chemists, biochemists, and biologists, and by the interactive generation of working hypotheses, their approval, modification, or falsification. In this sense, the close cooperation of theoretical and experimental scientists offers the best chances for success in drug research.

References

Akbaraly JP, Leng JJ, Bozler G, Seydel JK (1985) Quantitative relationship between trans-placental transfer and physicochemical properties of a series of heterogeneous drugs. In: Seydel JK (ed) QSAR and strategies in the design of bioactive compounds. Proceedings of the 5th European symposium on QSAR, Bad Segeberg, 1984. VCH, Weinheim, pp 313–317

Avdeef A (1992) pH-Metric log P. 1. Difference plots for determining ion-pair octanol-water partition coefficients of multiprotic substances, Quant Struct Act Relat 11:510–517

Avdeef A (1993) pH-Metric log P. 2. Refinement of partition coefficients and ionization constants of multiprotic substances. J Pharm Sci 82:183–190

Bartlett PA, Marlowe CK (1987) Evaluation of intrinsic binding energy from a hydrogen bonding group in an enzyme inhibitor. Science 235:569–571

Beddell CR (ed) (1992) The design of drugs to macromolecular targets. Wiley, Chichester

Bernard HEJ, Höffken HW, Hornberger W, Rübsamen K, Schmied B (1994) Thrombin-inhibiting decapeptides deduced from the C-terminus of hirudin. In: Hodges RS, Smith JA (eds) Peptides. Chemistry, structure and biology. Proceedings of the 13th American peptide symposium, 1993, Edmonton, Alberta, Canada. Escom Science, Leiden, pp 592–594

Blaney JM, Hansch C (1990) Application of molecular graphics to the analysis of macromolecular structures. In: Hansch C, Sammes PG, Taylor JB (eds) Quantitative drug design, Ramsden CA (ed). Comprehensive medicinal chemistry, vol 4. The rational design, mechanistic study and therapeutic application of chemical compounds. Pergamon, Oxford, pp 459–496

Böhm H-J (1992a) The computer program LUDI: a new simple method for the de novo design of enzyme inhibitors. J Comput Aided Mol Design 6:61–78

Böhm H-J (1992b) LUDI: Rule based automatic design of new substituents for enzyme inhibitor leads, J Comput Aided Mol Design 6:593–606

Böhm H-J (1993) Ligand design. In: Kubinyi H (ed) 3D QSAR in drug design. Theory, methods and applications. Escom Science, Leiden, pp 386–405

Böhm H-J (1994) The development of a simple empirical function to estimate the binding constant for a protein-ligand complex of known three-dimensional structure. J Comput Aided Mol Design 8:243–256

Borman S (1992) New 3-D search and de novo design techniques aid drug development. Chem Eng News (August 10):18–26

Boyd DB (1990) Successes of computer-assisted molecular design. In: Lipkowitz KB, Boyd DB (eds) Reviews in computational chemistry. VCH, New York, pp 355–371

Clozel J-P, Fischli W (1993) Discovery of remikiren as the first orally active renin inhibitor. Arzneimittelforschung (Drug Res) 43 (I):260–262

Convard T, Dubost, J-P, le Solleu H, Kummer E (1994) SmilogP: a program for a fast evaluation of theoretical log P from the Smiles code of a molecule. Quant Struct Act Relat 13:34–37

Dearden JC (1990) Molecular structure and drug transport. In: Hansch C, Sammes PG, Taylor JB (eds) Quantitative drug design, Ramsden CA (ed). Comprehensive medicinal chemistry, vol 4. The rational design, mechanistic study and therapeutic application of chemical compounds. Pergamon, Oxford, pp 375–411

Fujita T (1990) The extrathermodynamic approach to drug design. In: Hansch C, Sammes PG, Taylor JB (eds) Quantitative drug design, Ramsden CA (ed). Comprehensive medicinal chemistry, vol 4. The rational design, mechanistic study and therapeutic application of chemical compounds. Pergamon, Oxford, pp 497–560

Fujita T (1992) The role of QSAR in lead evolution. In: Kuchar M (ed) QSAR in design of bioactive compounds. Prous Science, Barcelona, pp 3–22

Ghose AK, Crippen GM (1986) Atomic physicochemical parameters for three-dimensional structure-directed quantitative structure-activity relationships I. Partition coefficients as a measure of hydrophobicity. J Comput Chem 7:565–577

Ghose AK, Crippen GM (1987) Atomic physicochemical parameters for three-dimensional structure-directed quantitative structure-activity relationships 2. Modeling dispersive and hydrophobic interactions. J Chem Inform Comput Sci 27:21–35

Gualtieri F, Teodori E, Bellucci C, Pesce E, Piacenza G (1985) SAR studies in the field of calcium(II) antagonists. Effect of modifications at the tetrasubstituted carbon of verapamil-like compounds. J Med Chem 28, 1621–1628

Gund P, Maggiora G, Snyder JP (1992) Industry leaders convene at Mackinac Island to discuss approaches for integrating CADD strategies into pharmaceutical R&D. Chem Design Autom News 7 (11):30–33

Hansch C, Hatheway GJ, Quinn FR, Greenberg N (1978) Antitumor 1-(X-aryl)-3,3-dialkyltriazenes. 2. On the role of correlation analysis in decision making in drug modification. Toxicity quantitative structure-activity relationships of 1-(X-phenyl)-3,3-dialkyltriazenes in mice. J Med Chem 21, 574–577

Hansch C, Leo A, Schmidt C, Jow PYC, Montgomery JA (1980) Antitumor structure-activity relationships. Nitrosoureas vs. L-1210 leukemia. J Med Chem 23:1095–1101

Hansch C, Li R-L, Blaney JM, Langridge R (1982) Comparison of the inhibition of Escherichia coli and Lactobacillus casei dihydrofolate reductase by 2,4-diamino-5-(substituted-benzyl)pyrimidines: quantitative structure-activity relationships, X-ray crystallography, and computer graphics in structure-activity analysis. J Med Chem 25:777–784

Hansch C, Björkroth JP, Leo A (1987) Hydrophobicity and central nervous system agents: on the principle of minimal hydrophobicity in drug design. J Pharm Sci 76:663–687

Hatheway GJ, Hansch C, Kim KH, Milstein SR, Schmidt CL, Smith RN, Quinn FR (1978) Antitumor 1-(X-aryl)-3,3-dialkyltriazenes. 1. Quantitative structure-activity relationships vs. L1210 leukemia in mice. J Med Chem 21:563–574

Höltje H-D (1982) Theoretische Untersuchungen zu Struktur-Wirkungsbeziehungen von ringsubstituierten Verapamil-Derivaten. Arch Pharm 315:317–323

Höltje H-D, Hense M (1985) Zur Struktur ringsubstituierter Verapamil-Derivate. Pharm Acta Helv 60:287–288

Höltje H-D, Kier LB (1974) A theoretical approach to structure-activity relationships of chloramphenicol and congeners. J Med Chem 17:814–819

Höltje H-D, Kier LB (1975) Nature of anionic or α-site of cholinesterase. J Pharm Sci 64:418–420

Kester WR, Matthews BW (1977) Crystallographic study of the binding of dipeptide inhibitors to thermolysin: implications for the mechanism of catalysis. Biochemistry 16:2506–2516

König H (1987) Calcium antagonists of verapamil type – chemical aspects. Actual Chim Ther 14:99–117

Kubinyi H (1978) Drug partitioning: relationships between forward and reverse rate constants and partition coefficient. J Pharm Sci 67:262–263

Kubinyi H (1979a) Lipophilicity and biological activity. Drug transport and drug distribution in model systems and in biological systems. Arzneimittelforschung (Drug Res) 29:1067–1080

Kubinyi H (1979b) Lipophilicity and drug activity. Fortschr Arzneimittelforsch (Prog Drug Res) 23:97–198

Kubinyi H (1988) Current problems in quantitative structure activity relationships. In: Jochum C, Hicks MG, Sunkel J (eds) Physical property prediction in organic chemistry. Springer, Berlin Heidelberg New York, pp 235–247

Kubinyi H (1993a) QSAR: Hansch analysis and related approaches. In: Mannhold R, Krogsgaard-Larsen P, Timmerman H (eds) Methods and principles in medicinal chemistry, vol 1. VCH, Weinheim

Kubinyi H (ed) (1993b) 3D QSAR in Drug Design. Theory, Methods and Applications, ESCOM Science, Leiden

Kubinyi H, Klebe G (1987) Structure activity relationships of calcium antagonists. In: Raviña E (ed) Actualidades de quimica terapeutica. Topics of current interest on medicinal chemistry, cursos e congresos da Universidade de Santiago de Compostela 55, Spain, pp 141–166

Leo AJ (1990) Methods of calculating partition coefficients. In: Hansch C, Sammes PG, Taylor JB (eds) Quantitative drug design, Ramsden CA (ed).

Comprehensive medicinal chemistry, vol 4. The rational design, mechanistic study and therapeutic application of chemical compounds. Pergamon, Oxford, pp 295–319

Marshall GR (1993) Binding-site modeling of unknown receptors. In: Kubinyi H (ed) 3D QSAR in drug design. Theory, methods and applications. Escom Science, Leiden, pp 80–116

Marshall GR, Naylor CB (1990) Use of molecular graphics for structural analysis of small molecules. In: Hansch C, Sammes PG, Taylor JB (eds) Quantitative drug design, Ramsden CA (ed). Comprehensive medicinal chemistry, vol 4. The rational design, mechanistic study and therapeutic application of chemical compounds. Pergamon, Oxford, pp 431–458

Montgomery JA, Niwas S (1993) Structure-based drug design. Chemtech 23 (November):30–37

Montgomery JA, Niwas S, Rose JD, Secrist JA III, Babu YS, Bugg CE, Erion MD, Guida WC, Ealick SE (1993) Structure-based design of inhibitors of purine nucleoside phosphorylase. 1. 9-(Arylmethyl) derivatives of 9-deazaguanine. J Med Chem 36:55–69

Morgan BP, Scholtz JM, Ballinger MD, Zipkin ID, Bartlett PA (1991) Differential binding energy: a detailed evaluation of the influence of hydrogen-bonding and hydrophobic groups on the inhibition of thermolysin by phosphorus-containing inhibitors. J Am Chem Soc 113:297–307

Navia MA, Murcko MA (1992) Use of structural information in drug design. Curr Opin Struct Biol 2:202–210

Parker EM, Grisel DA, Iben LG, Shapiro RA (1993) A single amino acid difference accounts for the pharmacological distinctions between the rat and human 5-hydroxytryptamine 1B receptors. J Neurochem 60:380–383

Reich SH, Webber SE (1993) Structure-based drug design (SBDD): every structure tells a story. Perspect Drug Discov Design 1:371–390

Rekker RF (1977) The hydrophobic fragmental constant. Its derivation and application. A means of characterizing membrane systems, pharmacochemical library 1. Elsevier, Amsterdam

Rekker RF, Mannhold R (1992) Calculation of drug lipophilicity. The hydrophobic fragmental constant approach. VCH, Weinheim

Rydel TJ, Ravichandran KG, Tulinsky A, Bode W, Huber R, Roitsch C, Fenton JW II (1990) The structure of a complex of recombinant hirudin and human α-thrombin. Science 249:277–280

Rydel TJ, Tulinsky A, Bode W, Huber R (1991) Refined structure of the hirudin-thrombin complex. J Mol Biol 221:583–601

Scherrer RA, Howard SM (1977) Use of distribution coefficients in quantitative structure-activity relationships. J Med Chem 20:53–58

Seidel W, Meyer H, Born L, Kazda S, Dompert W (1985) Rigid calcium antagonists of the nifedipine-type: geometric requirements for the dihydropyridine receptor. In: Seydel JK (ed) QSAR and strategies in the design of bioactive compounds. Proceedings of the 5th European symposium on QSAR, Bad Segeberg, 1984. VCH, Weinheim, pp 366–369

Selassie CD, Klein TE (1993) Building bridges: QSAR and molecular graphics. In: Kubinyi H (ed) 3D QSAR in drug design. Theory, methods and applications. Escom Science, Leiden, pp 257–275

Silipo C, Vittoria A (1979) A reanalysis of the structure-activity relationships of sulfonamide derivatives. Farmaco Ed Sci 34:858–868

Sussman JL, Harel M, Frolow F, Oefner C, Goldman A, Toker L, Silman I (1991) Atomic structure of acetylcholinesterase from Torpedo californica: a prototypic acetylcholine-binding protein. Science 253:872–879

Suzuki T (1991) Development of an automatic estimation system for both the partition coefficient and aqueous solubility. J Comput Aided Mol Design 5:149–166

Suzuki T, Kudo Y (1990) Automatic log P estimation based on combined additive modeling methods. J Comput Aided Mol Design 4:155–198

Timmermans PBMWM, de Jonge A, van Meel JCA, Slothorst-Grisdijk FP, Lam E, van Zwieten PA (1981) Characterization of α-adrenoceptor populations. Quantitative relationships between cardiovascular effects initiated at central and peripheral α-adrenoceptors. J Med Chem 24:502–507

Timmermans PBMWM, de Jonge A, Thoolen MJMC, Wilffert B, Batink H, van Zwieten PA (1984) Quantitative relationships between α-adrenergic activity and binding affinity of α-adrenoceptor agonists and antagonists. J Med Chem 27:495–503

Topliss JG (1993) Some observations on classical QSAR. Perspect Drug Discov Design 1:253–268

10 The Role of Structure-Based Ligand Design in Industrial Pharmaceutical Research

J. P. Tollenaere

10.1 Introduction

Nowadays, new and better medicines result more often than not from a concerted multidisciplinary approach towards treating a medical problem or curing a disease. Modern drug research has to be a multidisciplinary endeavor in which medicinal chemists, pharmacologists, molecular biologists, biochemists, pharmacists, computer scientists, theorists, and clinicians work together towards the common goal of finding and developing an improved or new medicine.

Modern pharmaceutical research, however, faces fundamental obstacles bearing on yet ill-understood complexities of living matter. Scientific and technological developments in a wide range of fields, such as molecular biology, physiology, chemistry, physics, information and computer science, have facilitated the unraveling of these complexities. With our better understanding of normal and diseased states, our desire to intervene pharmacologically in disease processes has grown.

The need for continuing development of new drugs hardly needs to be pointed out, in view of the global health and disease situation. The need for drugs that will effectively halt the spread of the AIDS virus is painfully apparent. Other scourges such as tuberculosis, malaria, and various parasitic plagues continue to afflict many millions of people. With increased life expectancy, diseases such as Alzheimer's and other central nervous system malfunctions will increasingly require the development of new drugs.

Traditionally, the drug discovery process relied almost exclusively on the mere synthesis and subsequent pharmacological testing of many thousands of chemical compounds. Although almost from the very beginning of contemporary medicinal chemistry the idea of a relation between chemical properties and biological activity was accepted, the synthesis of compounds was to a large extent dictated by what was synthetically feasible. Pharmacological testing was based on relatively simple animal models. By its very nature, the entire process of drug discovery was not as multidisciplinary and integrated as it is today in the modern pharmaceutical industry. Yet, it is an undeniable fact that the traditional approach has led to the discovery of most prototype structures and to the development of most drugs in use today. In this chapter, we will express our views on the position and the role of theoretical medicinal chemistry in the multidisciplinary and collaborative endeavor in which a large diversity of scientific disciplines is brought together with the objective of discovery and development of improved and/or new drugs. In particular, the value and usefulness of structure-based ligand design will be discussed in relation to the overall development of a new drug.

10.2 The Position of the Theoretical Medicinal Chemist

As shown in Fig. 1, the process of putting a new drug into the hands of the medical practitioner or a pharmacist is laborious and lengthy. Although admittedly rough and very schematic as depicted here, the flow of events usually starts from an idea triggered by scanning and reading of the literature. A chemist may be led into devising the synthetic pathway of a new chemical, or the pharmacologist may come up with setting up a new assay. Reading the patent literature may assure the chemist that his new compound is really novel and patentable. The flow of events may also be initiated by a clinician pointing out the need for a particular medicine. Sometimes mere serendipity may lie at the origin of the sequence of events.

Once it has been decided to embark on the synthesis of a series of compounds, all more or less related to a parent structure, the new

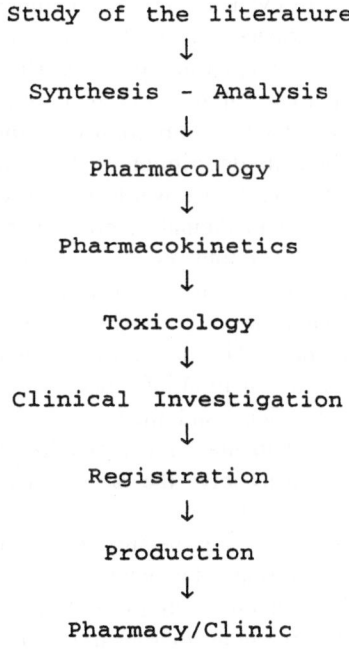

Study of the literature
↓
Synthesis - Analysis
↓
Pharmacology
↓
Pharmacokinetics
↓
Toxicology
↓
Clinical Investigation
↓
Registration
↓
Production
↓
Pharmacy/Clinic

Fig. 1. The position of the theoretical medicinal chemist

chemical entities are analyzed by, for example, nuclear magnetic resonance (NMR), mass spectrometry, ultraviolet (UV) light, or infrared (IR)
light. Basic chemical properties such as chemical composition, purity,
solubility, partition coefficient, optical rotation, and pK values may be
measured.

Those compounds meeting a given purity standard may then be
forwarded to the pharmacologists who perform simple and quick tests
that can characterize the compounds at a given standard dose as having,
for example, anticholinergic, antihistaminic, dopaminergic, or analgesic
activity or having no activity at all. The more interesting compounds
can subsequently be evaluated in more sophisticated pharmacological
models or in receptor binding assays.

The few tens of compounds surviving all these tests are by then
relatively well characterized in terms of their biological properties and
can be evaluated from the pharmacokinetic and toxicological point of
view. The most promising compounds surviving this further round of
testing can then finally be evaluated in several clinical trials.

All the chemical, pharmacological, pharmacokinetic, toxicological,
and clinical data collected over a time span of 8–10 years form the basis
for the registration file to be submitted to the public health authorities,
who may finally approve the new drug for use in clinical practice.

From this very schematic description of the flow of events, or the
trajectory a molecule has to follow towards the status of drug, it must be
apparent that the theoretical medicinal chemist is to be found right at the
beginning of this time-consuming process. In particular, he or she is
typically to be found in the library if not in the organic synthetic or
physical chemistry laboratories or discussing biological data with pharmacologists or biochemists. The theoretical medicinal chemist deals
with molecules for which structural information is to be obtained by a
variety of chemical, physical, and theoretical techniques. Different
techniques provide complementary types of information that together
can be employed to help to solve the basic problem: the biological
response is a function of structure.

In the following, an overview is presented of the tools and techniques
used and some issues are discussed which bear on the real contributions
a theoretical medicinal chemist can make towards a better understanding of how known drug molecules act or how the discovery of new
ones may be facilitated.

10.3 The Basic Assumptions

The basic assumption when dealing with molecules that elicit a biological response (BR) is that there is relation between the structure (S) of the molecule and the physicochemical properties of the molecules that are ultimately responsible for the biological activity.

Succinctly stated: BR=f(S), where S may be any or a combination of some of the properties listed below.

- Lipophilicity (log P), solubility
- Ionization constant (pK)
- Shape (volume, conformation, configuration, solvent accessible surface area)
- Atomic charges, dipole moment (μ)
- Molar refraction, polarizability
- Electronic structure, energy levels, e.g., HOMO, LUMO, reactivity indices, rotational barriers, entropic and hydrophobic effects, hydrogen bonds.

This list, though by no means exhaustive, contains a number of properties that are only accessible by experiment, such as log P and pK, while others can only be obtained by computational means. The choice as to which are possibly correlated with the biological activity is not easy and often depends on trial and error, particularly when the underlying reaction mechanism is not known. Therefore, it is of interest to envisage the hypothetical sequence of events during the trajectory of a drug or ligand from its injection into the biological system until it elicits a biological effect (Fig. 2).

| Partitioning | Desolvation | Interaction | Biological |
| Distribution | Recognition | "receptorization" | Response |

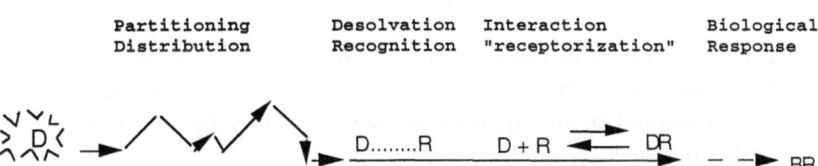

Fig. 2. Trajectory of a drug (*D*) as from the injection into the biological system until its interaction with the receptor (*R*), ultimately leading to the biological response (*BR*)

At the stage of partitioning and distribution one might envisage a drug or ligand molecule crossing several cell membranes and lipid barriers and that properties such as partitioning behavior, as reflected in the log P parameter, may satisfactorily describe the underlying phenomena. At the desolvation and recognition phase, physical quantities and concepts such as hydrogen bond formation and breaking and entropic and hydrophobic effects may be invoked to rationalize biological data. At the drug–receptor interaction stage, physical properties such as conformation, configuration, dipolar effects, the highest occupied molecular orbital (HOMO) or lowest unoccupied molecular orbital (LUMO) may effectively govern the interaction. In practice, however, there are no sharp divisions between the various stages. Thus, the appearance of log P in a regression equation may, in fact, reflect a genuine partitioning effect governing biological activity, as might be expected in in vivo testing, but may equally well signal the importance of desolvation prior to binding to the hydrophobic surface of a protein.

The conformation of a ligand or drug is of utmost importance in all quantitative structure–activity relationships (QSAR), except in the traditional Hansch-type QSAR where only topological or two-dimensional (2D) information of a congeneric series of molecules is needed (Fujita 1990). In fact, many, if not all, physicochemical properties associated with structure S depend on the spatial arrangement of the atoms in the molecule. Therefore, conformational aspects are core concepts in any modeling "experiment." A good understanding of the limitations and applicability of the techniques used to determine the conformation of molecular structures is of crucial importance in structure-based ligand design.

10.4 Conformational Analysis

At the forefront of any molecular modeling experiment is the question as to what shape(s) or conformation(s) a molecule can adopt. Molecules can be observed in three aggregation states, i.e., the solid, the liquid, and the gaseous state.

Solid State. Provided suitable crystals are available, X-ray diffraction experiments lead to the precise location of each atom of the molecule within the crystal lattice. Although the conformation observed in the crystal possibly pertains to a minimum energy conformation (MEC), there is no a priori reason for this conformation to be biologically relevant. That is, though experimentally determined, this conformation is not necessarily the conformation recognized by the receptor or the conformation required for a productive drug–receptor interaction.

Liquid State. Molecules in solution form the second aggregation state. The method of choice for determining a molecule's conformation in solution is NMR. Once more, although drug molecules in physiological conditions are in the dissolved state, the solvents used in an NMR experiment, such as chloroform or dimethylsulfoxide (DMSO), must be considered to be poor mimics of physiological environmental conditions.

Gaseous State. Molecules can also be studied in a third aggregation state, namely, the gaseous or isolated state. Apart from microwave spectroscopic techniques suitable for the analysis of relatively small molecules, the method of choice for studying the conformational aspects of drug molecules is computation, either by quantum chemical or molecular mechanical procedures. Calculations in the isolated state or in vacuo calculations are usually conducted in the complete absence of any environmental effects of the medium surrounding a molecule and therefore their biological relevance may be questioned.

Table 1 summarizes the benefits and disadvantages of the three methods used for the determination of the conformation of a molecule. Despite the fundamental objections that can be raised against any of these approaches, the combination of the three methods yields a complete picture of the conformational aspects of a molecule. Armed with this knowledge, one may try to propose a pharmacophore (i.e., the spatial disposition of atoms or groups of atoms that is required for the recognition of and interaction with a receptor) that can either be used as a lead for the synthesis of new compounds or be employed to construct a hypothetical receptor model.

As a molecule becomes conformationally more flexible one is rapidly faced with the problem of finding the MEC. The magnitude of the problem can be illustrated by consideration of a molecule with, for example, six rotatable bonds. If it is assumed that each bond can rotate

Table 1. Methods of conformational analysis

Method	Environment	Information
X-ray diffraction		
Crystalline state	Like molecules	Atomic coordinates
	Solvent molecules	One conformation
	Counter ions	in most cases
NMR		
Solution	Like molecules	In many cases partial
	Solvent molecules	answer conformation
Computation		
Isolated state	None unless solvent	All conformations
	model is included	Electronic structure
		Energy levels

NMR, nuclear magnetic resonance

a full circle in steps of $10°$, one faces the combinatorial problem of calculating N_{conf} conformations (Eq. 1):

$$N_{conf} = \left(\frac{360}{10} \right)^n = 36^6 \approx 2.1 \times 10^9 \tag{1}$$

At a speed of 100 conformation evaluations per second, this brute force approach means asking for some 250 days of CPU time from your local computer department! It is clear that the brute force approach using a traditional sequential-architecture computer is only feasible for molecules with four to five rotatable bonds.

Instead of using the systematic search in torsional angle space, also called the nonadiabatic grid search, one may resort to the various energy minimization techniques (Leach 1991; Howard and Kollman 1988). All standard energy minimization methods, such as steepest descents, conjugate gradients, and Newton-Raphson, always proceed downhill. Thus, the energy and the corresponding structure obtained after the minimization is strongly dependent on the starting structure and does not necessarily represent the MEC. Sampling the conformational hypersurface by random generation of starting structures followed by static energy minimization may lead to the identification of the MEC. Molecular dynamics (MD) (Howard and Kollman 1988), simulated annealing (Wilson et al. 1991) and Monte Carlo methods (Nayeem et al. 1991) are suitable for the generation of random starting structures.

10.5 Computational Techniques

Broadly speaking, there are two fundamental classes of computational methodologies that can be used to simulate chemical behavior: the quantum chemical and the molecular mechanical approaches, each having its advantages and disadvantages. In general, quantum chemical calculations, even at the semi-empirical level of approximation, are not practically feasible for molecular systems containing more than 200 to 300 atoms. For systems having up to several thousands of atoms molecular mechanics methodologies are invariably used.

10.5.1 Quantum Chemical Calculations

For relatively small molecular systems, quantum chemical calculations ranging from the semi-empirical approximation up to the highest quality ab initio methods are employed. Again, the choice of a particular approximation depends on the system at hand. Because even quantum chemical methodologies have their strong and weak points, only experience with the various methods can tell which method is to be used for a particular problem. A striking example of the varying results obtained

Table 2. Charge on the N atom of amines, from various semi-empirical quantum chemical methods

Compound	PM3	AM1	CNDO	MNDO	STO-3G
NH_3	0.007	−0.396	−0.2457	−0.2878	−0.481
CH_3NH_2	−0.030	−0.352	−0.203	−0.3098	−0.4122
$(CH_3)_2NH$	−0.057	−0.308	−0.167	−0.3331	−0.3479
$(CH_3)_3N$	−0.072	−0.266	−0.1370	−0.3573	−0.2876

PM3, parametrized model 3; AM1, Austin model 1; CNDO, complete neglect of differential overlap; MNDO, modified neglect of differential overlap; STO-3G, Slater type orbital – 3 gaussians by the various semi-empirical methods compared with STO-3G results is shown in Table 2.

It is clearly seen that the trend of the inductive effect of methylation of the N atom can lead to opposite results, depending on the particular method used. In fact, Austin Model 1 (AM1) and complete neglect of differential overlap (CNDO) predict a progressively more negatively charged N in the series: NH_3, primary, secondary, and tertiary amine, whereas Modified Neglect of Differential Overlap (MNDO) and Parametrized Model 3 (PM3) predict the opposite.

Despite these cautionary remarks, it must be stated that semi-empirical calculations prove to be of major value in the fairly accurate description of molecular systems and in predicting chemical behavior in a series of compounds (Zerner 1991).

10.5.2 Molecular Mechanics

The purely empirical approach or the so-called force field calculation is based on a purely mechanical model of a molecule, which is considered to be an assembly of point masses connected by springs and susceptible to classical motions such as bond stretching, bond angle bending, torsional motions, and nonbonded Van der Waals and electrostatic interactions (Siebel and Kollman 1990). Owing to the simplicity of the mathematical expression (Eqs. 2,3), molecular mechanics (MM) are particularly well suited for the calculation of the internal energy V of a molecular system comprising several thousands of atoms.

$$V = \sum V_{bonds} + \sum V_{angles} + \sum V_{torsion} +$$
$$\sum V_{vdw} + \sum V_{coul} \tag{2}$$

The MM approach is widely used for the elucidation of the conformational aspects of both small drug molecules and macromolecular structures. Due to their empirical nature, MM calculations are often hampered by the lack of suitable parameters K_r, K_q, K_f, a_{ij} and b_{ij}, which are often not available for particular molecular fragments (Bowen and Allinger 1991). Due to its classical nature, MM is totally inadequate for model bond breaking and bond formation processes.

$$V = \sum_{\text{bonds}} K_r (r - r_{eq})^2 +$$

$$\sum_{\text{angles}} K_\theta (\theta - \theta_{eq})^2 +$$

$$\sum_{\text{tors}} \frac{1}{2} K_\phi [1 + \cos(n\phi - \gamma)] + \qquad (3)$$

$$\sum_{i<j} \frac{b_{ij}}{r_{ij}^{12}} - \frac{a_{ij}}{r_{ij}^{6}} +$$

$$\sum_{i<j} \frac{1}{\varepsilon} \frac{q_i q_j}{r_{ij}}$$

10.5.3 Molecular Dynamics

Instead of the static picture one gets from the computations thus far discussed, the atoms of a molecule are actually in constant motion. These motions around an equilibrium position, or even larger fluctuations involving the movement of side chains of the amino acids of proteins, can be simulated by the technique of MD. MD simulations are based on the knowledge of the energy of a system (Eq. 2) as a function of the atomic coordinates (van Gunsteren and Berendsen 1990; Brunger and Karplus 1991). The force (F_i) acting on each atom is related to the first derivative of the potential energy V with respect to the atom position. Solving Newton's equation by using this force leads to the motion of the atoms as a function of time (Eq. 4):

$$F_i = -\frac{\partial V_i}{\partial x_i} \qquad F_i = m_i a_i \qquad (4)$$

MD simulations are frequently used to examine the possible conformational domains of small molecules and macromolecular systems and may lead to the assessment of the conformational flexibility of a molecule. MD calculations are one of the strategies that can be used to generate low energy conformations.

10.6 Data Bases of Molecular Structure

Molecular structure databases combined with computer-assisted molecular modeling are of vital importance in ligand design. As already mentioned, X-ray crystallography is one of main sources of information bearing on the 3D structure of small molecules and macromolecules. It is clear that molecular modeling without the knowledge gained from X-ray crystallography would have to rely solely on theoretical models of molecular structure. There are two X-ray crystallographic data bases that are used by those interested in structure-based ligand design.

The Cambridge Structural Database (CSD), which contains the X-ray structure coordinates of small organic and organometallic compounds (120481 entries in the April 1994 release), is the prime source of experimentally determined information regarding the 3D characteristics of small organic molecules (Allen and Davies 1988). In many instances retrieving structures from the CSD will give high-quality models of structures, which can subsequently be modified into the desired structures by use of molecular modeling techniques. Often, X-ray structures are used as input structure for theoretical calculations.

The Brookhaven Protein Data Bank (PDB) contains the coordinates of protein and nucleic acid structures (Bernstein et al. 1977). The PDB (2327 entries in the January 1994 release) offers a rich source of information about the tertiary structure of proteins. Detailed analysis of the data often promotes a better understanding of specific molecular characteristics of ligand–protein interactions and interaction sites.

Pharmaceutical companies typically maintain data bases of the compounds that have been synthesized over the years. These data bases, often containing several hundreds of thousands of compounds, store 2D information of chemical structures. With the advent of software that can generate 3D structures from 2D information, a new wealth of 3D information becomes available for 3D searching and for identifying pharmacophoric patterns of functional groups or atoms important for recognition of and interaction with a receptor, leading to a given biological response. Three-dimensional searching, or data mining, is becoming an important tool in lead generation (Martin et al. 1991; Sheridan et al. 1989). In particular, 3D searching seems to be particularly useful in "reviving" the older structures of a corporate data base. In fact, it often happens that in a given current research project, medicinal chemists tend to concentrate on the structures they are currently dealing with, and their analogs, while forgetting that older structures that have not been evaluated in the pharmacological tests of today could equally well be candidate compounds for the current research project.

In general, data bases containing either experimentally derived structures, such as CSD and PDB, or calculated structures are indispensable tools for theoretical or computational medicinal chemists. Fast and easy access to 3D structural information is of crucial importance for molecular modeling. It offers a wealth of information regarding inter- and intramolecular architecture and may provide suitable template structures that can then be used for further in computro modification.

10.7 Molecular Mechanics

The potential energy of a molecular system consisting of interacting particles should, in principle, be treated by use of quantum mechanical methods. As the size of biological systems, and thus the number of particles, is so large, quantum mechanical methods are in practice not feasible for the description of peptides or proteins. One could argue that semi-empirical quantum mechanical methods are feasible for relatively small molecules. Although these methods are indeed frequently and successfully used for small molecules, it remains also true that it is not an easy matter to combine quantum mechanical calculations with MM calculations if one wants to simulate and model the interaction characteristics and energetics of a ligand (the small molecule) with its receptor (the protein molecule). Therefore, there is certainly a need for a theoretical formalism capable of treating large and small molecules simultaneously.

Most calculations on macromolecular structures are carried out using empirical potential functions (Eqs. 2,3). These are simple analytical functions expressing the potential energy of a molecule in terms of valence interactions. The reliability of empirical potential energy or molecular mechanics calculations depends on the energy terms included in the total energy function and the numerical values of the parameters. As MM calculations can routinely be done by virtually all modeling software packages, it is of more than academic interest to have a closer look into the applicability and limitations of force fields.

The advantage of being conceptually simple, because valence force fields are expressed in terms of bond stretching, bond angles, torsion angles, etc., is at the same time a weak point. Over the years, several force fields have evolved and gradually migrated from academic to commercial environments. By now, every commercially available molecular modeling software vendor has its own force field in different stages of development. In fact, companies that started offering small molecule modeling capability have acquired and incorporated an academic force field, in order to do macromolecule modeling.

Similarly, modeling software originally intended for macromolecules gradually saw the incorporation of force fields capable of treating small organic molecules. Not only have existing force fields different ranges of applicability, but they also have different mathematical functions for the expression of the potential energy. Force fields may, for example, differ in how nonbonded van der Waals interactions are treated; they may differ in the way hydrogen bond interactions are taken into account; some are purely diagonal while others may include off-diagonal cross terms; the way electrostatic interactions are treated may vary widely, ranging from simply omitting electrostatics to the use of point charges, bond dipoles, atomic polarizability, and higher-order atomic multipole moments (Brooks et al. 1983; Weiner et al. 1984; Clark et al. 1989; Dinur and Hagler 1991). Adding to the despair of the user, software vendors acknowledge the less-than-optimal quality of their force fields and therefore, more often than not, each new release of a modeling package contains some changes (not always fully documented) in the force field used. The effort to try to improve or alter things, laudable as it is in principle, has, in practice, the effect that customers have to invest much time in bug hunting and in evaluating the new "improvements." Furthermore, research projects lasting considerably longer than the software release cycle time may not always profit from the latest release, because results of the computations differ if the force field is altered from one release to the other. Keeping older releases alive along side the latest release may help in ensuring the continuity and consistency of some research projects.

In general, where it is even obvious that successive releases of one and the same software package are not always transferable, force fields from different origins are not transferable. Force fields that are accurate over a limited range of compounds are not necessarily accurate for a somewhat different set of compounds. As long as force fields continue to contain parameters that are not truly transferable, the situation will prevail in which the ultimate choice of a force field is to a large extent a software-vendor-driven process. If a pharmaceutical company already has a more or less well established software package, it is not easy to convince it to break with the past and acquire a new package (and possibly another force field) or to add another package to the existing one(s).

In conclusion, due to their importance in the simulation of the chemical behavior of small molecules and macromolecular systems, the accuracy and the applicability of the routinely used force fields are and should be issues of constant concern for the theoretical medicinal chemist using them.

10.8 The Role of Computational Chemistry in Drug Discovery

In this contribution we have presented some topics and issues related to structure-based ligand design. Admittedly, the number of topics discussed is by no means exhaustive and is mainly inspired by personal bias and daily practice over the last 25 years. Nevertheless, it should be clear that, due to the position of the theoretical medicinal chemist in the long process of original concept towards the point of making a drug available to a physician and his or her patients, computational chemists are dealing in the vast majority of cases with ligands. From this point of view, "computer-assisted drug design" is an utter misnomer. Also the expression "rational drug design," which falsely may imply that drugs can be designed like pieces of furniture and that the historically successful screening approach was irrational, has to be averted.

Computer-assisted molecular modeling (CAMM) uses the methods and techniques of computational chemistry to describe and possibly predict physicochemical properties of molecules or ligands and to simulate their behavior along their itinerary from their injection into the biological system towards recognition of and interaction with a macromolecular receptor molecule.

It is beyond any doubt that computational chemistry has made great strides in achieving satisfactory agreement between computed and ex-

perimentally determined properties of molecules. Even a casual brows-
ing through the current chemical literature should convince anyone that
molecular properties such as 3D structure, energies, molecular interactions, and spectroscopic properties are amenable to successful computation. Applying the methods and techniques of computational chemistry
to ligands or biologically active molecules does not warrant the notion
that one is designing drugs: it simply means that one is calculating
molecular properties that possibly lie at the basis of biological activity.
As such, computational chemistry may be used to find useful correlations between chemical properties and biological activity, thereby providing a rationale as to why some compounds are biologically active
and some are not. At this stage, theoretical medicinal chemists can
influence the direction a research project may take.

A more direct contribution from theoretical medicinal chemists can
be expected if the 3D structure of the binding site of the receptor is
known. In this case, a more detailed description of the energetics and
the 3D characteristics of the ligand–receptor complex can be computed.
These types of computational experiments may result in a better understanding of how and why a ligand binds to a receptor, in terms of its
structural characteristics, and may lead to suggestions of the synthesis
of new analogs.

Frequently, the 3D structure of the target is not known and then one
can resort to homology modeling (Chothia and Lesk 1986) to deduce
the 3D structure of a protein if its amino acid sequence is known and
when structural data of a homologous protein are available. Other
strategies include receptor mapping and de novo ligand design (Böhm
1992, 1994; Leach and Kilvington 1994), whereby possible interaction
sites are identified and molecules that are complementary to these
interaction sites are searched in 3D data bases.

If theoretical chemists really want to have a significant impact on the
daily life of experimental chemists or biochemists, the quantity of
interest is the free energy, ΔG, of a system. All chemical behavior is
determined by differences in ΔG between reactants and products in a
reaction, or between reactants and the transition state. Thus, the free
energy of binding $\Delta G = \Delta H - T\Delta S$ has contributions from the enthalpy
ΔH and the entropy ΔS. As ΔG is a state function that depends on the
extent of phase or configuration space accessible to the molecular
system, the computation of ΔG of a molecular system is virtually

impossible. However, the relative ΔG of a molecular association between two structurally closely related ligands can be approached on the basis of the so-called thermodynamic cycle and free energy perturbation (FEP) methods (Kollman 1993). The implementation of FEP algorithms in molecular modeling software packages, combined with the ever-increasing development and availability of faster computer hardware usher in a new era in which entropic and solvent effects may be properly taken into account.

10.9 Concluding Remarks

In this contribution some apects of computational chemistry pertaining to structure-based-ligand design have been briefly discussed. An attempt has been made to delineate the role of computational techniques that can be used to characterize the chemical properties and behavior of molecular systems. As such, the use of computational methods only produces physical quantities and numbers. The connotation of (biologically active) ligand design comes in when these physical quantities are related and used to rationalize biological data. It should be stressed that the results of structure-based ligand design are not clear-cut recipes useful for further action by organic synthetic chemists or pharmacologists as is often seen in computer-assisted manufacturing. In fact, as has been stated in the introduction, modern pharmaceutical research is limited by the complexities of the biological processes and by the same token the theoretical chemist who wants to relate these biological processes to the physical properties of the ligands and receptors presumably responsible or involved in drug action. Therefore, it should be realized that because of these staggering complexities, awesome simplifications and approximations must be used. On the one hand, owing to the sheer size of macromolecular structures in general, their energetics and 3D aspects can only be approached by molecular mechanics, thereby excluding, for example, bond breaking and bond formation processes. On the other hand, the complexities of the processes playing a role in living matter, which pre-eminently is a dissipative system, force one to construct and study simplified models of biochemical reality.

Nevertheless, if ligand–receptor recognition and interaction are considered to be necessary, but not sufficient steps for biological activity, the theoretical chemical approach to the problem, despite all its approximations, does make significant contributions to structure-based ligand design (Lam et al. 1994). It is our experience that computer-assisted molecular modeling is an indispensable tool for displaying and manipulating molecular structures generated by experimental and/or theoretical techniques. In fact, the mere viewing and manipulation of 3D models of small molecules on a computer graphics screen gives synthetic organic chemists a better understanding of their current molecules and quite frequently offer them some clues as to what else they could synthesize. Likewise, it appears that biochemists profit from "seeing" and "looking" into their target structure. In general, molecular modeling definitely stimulates the creativity of those involved in the study and analysis of biologically active ligands.

It should be pointed out, however, that whatever the degree of sophistication of the employed hardware and software, structure-based ligand design is but a small step in the arduous and costly process from concept towards a useful medicine.

References

Allen FH, Davies JE (1988) File structures and search strategies for the Cambridge structural database. Crystallogr Comput 2:271–289

Bernstein FC, Koetzle TF, Williams GJB, Meyer EF, Brice MD, Rodgers JR, Kennard O, Shimanouchi T, Tasumi M (1977) The Protein Data Bank: a computer-based archival file for macromolecular structures. J Mol Biol 112:535–542

Böhm H-J (1992) The computer program LUDI: a new method for the de novo design of enzyme inhibitors. J Comput Aided Mol Design 6:61–78

Böhm H-J (1994) The development of a simple empirical scoring function to estimate the binding constant for a protein-ligand complex of known three-dimensional structure. J Comput Aided Mol. Design 8:243–256

Bowen JP, Allinger NL (1991) Molecular mechanics: the art and science of parametrization. In: Lipkowitz KB, Boyd DB (eds) Reviews in computational chemistry, vol 2. VCH, Weinheim, pp 81–97

Brooks BR, Bruccoleri ER, Olafson ER, States DJ, Swaminathan S, Karplus M (1983) CHARMm: a program for macromolecular energy minimization and dynamic calculations. J Comput Chem 4:187–217

Brunger AT, Karplus M (1991) Molecular dynamics simulations with experimental restraints. Accid Chem Res 24:54–61

Chothia C, Lesk AM (1986) The relation between the divergence of sequence and structure in proteins. EMBO J 5:823–826

Clark M, Cramer RD III, Van Opdenbosch N (1989) Validation of the general purpose tripos 5.2 force field. J Comput Chem 10:982–1012

Dinur U, Hagler AT (1991) New approaches to empirical force fields. In: Lipkowitz KB, Boyd DB (eds) Reviews in computational chemistry, vol 2. VCH, Weinheim, pp 99–164

Fujita T (1990) The extrathermodynamic approach to drug design. In: Hansch C, Sammes PG, Taylor JB, Ramdsen CA (eds) Comprehensive medicinal chemistry. Quant Drug Design 4:497–560

Howard AE, Kollman PA (1988) An analysis of the current methodologies for conformational searching of complex molecules. J Med Chem 31:1669–1675

Kollman P (1993) Free energy calculations: applications to chemical and biochemical phenomena. Chem Rev 93:2395–2417

Lam PYS, Jadhav PK, Eyermann CJ, Hodge CN, Ru Y, Bacheler LT, Meek JL, Otto MJ, Rayner MM, Wong YN, Chang C-J, Weber PC, Jackson DA, Sharpe TR, Erickson-Viitanen S (1994) Rational design of potent, bioavailable, nonpeptide cylic ureas as HIV protease inhibitors. Science 263:380–384

Leach AR (1991) A survey of methods for searching the conformational space of small and medium-sized molecules. In: Lipkowitz KB, Boyd DB (eds) Reviews in computational chemistry, vol 2. VCH, Weinheim, pp 1–55

Leach AR, Kilvington SR (1994) Automated molecular design: a new fragment-joining algorithm. J Comput Aided Mol Design 8:283–298

Martin YC, Bures MG, Willet P (1991) Searching databases of three-dimensional structures. In: Lipkowitz KB, Boyd DB (eds) Reviews in computational chemistry, vol 1. VCH, Weinheim, pp 213–263

Nayeem A, Villa J, Scheraga HA (1991) A comparative study of simulated-annealing and Monte Carlo with minimization approaches to the minimum-energy structures of polypeptides: [met]-enkephalin. J Comput Chem 12:594–605

Sheridan RP, Rushinko A III, Nilakantan R, Venkataraghavan R (1989) Searching for pharmacophores in large coordinate data bases and its use in drug design. Proc Natl Acad Sci USA 86:8165–8169

Siebel GL, Kollman PA (1990) Molecular mechanics and the modelling of drug structures. In: Hansch C, Sammes PG, Taylor JB, Ramdsen CA (eds) Comprehensive medicinal chemistry. Quant Drug Design 4:125–138

van Gunsteren WF, Berendsen HJC (1990) Computer simulation of molecular dynamics: methodology, applications and perspectives in chemistry. Angew Chem 29:992–1023

Weiner SJ, Kollman PA, Case DA, Singh UC, Ghio C, Alagona G, Profeta S
    Jr, Weiner P (1984) A new force field for molecular mechanical simulation
    of nucleic acids and proteins. J Am Chem Soc 106:765–784
Wilson SR, Cui W, Moskowitz JM, Schmidt KE (1991) Applications of simu-
    lated annealing to the conformational analysis of flexible molecules. J
    Comput Chem 12:342–349
Zerner MC (1991) Semiempirical molecular orbital methods. In: Lipkowitz
    KB, Boyd DB (eds) Reviews computational chemistry, vol 2. VCH, Wein-
    heim, pp 313–365

11 Drug Design Methods in Real-Life Situations: Recent Examples and Future Opportunities

R. M. Hyde

11.1 Introduction

Computer-aided drug design (CADD) methods have been developing rapidly and promise to continue doing so. The pharmaceutical research environment is becoming increasing challenging. Researchers are required to identify high quality molecules to hit challenging therapeutic targets and to do it on a shorter time scale than ever.

In this environment it is clearly important that any opportunities for greater effectiveness offered by CADD are fully exploited. It is also crucial that the methods being developed are those which are required by the industrial researchers to do their job.

An effective partnership has evolved between the experience of the medicinal chemist and the opportunities offered by the CADD methods.

Recently, there have been important advances in technology which are having a major impact in our modus operandi.

In this presentation, a brief overview will be given of parts of the drug discovery process in which changes in technology together with changes in perception have led to changes in practice. A priority has been to get the best out of the new developments and ensure a productive complementarity between apparently competitive approaches.

CADD is a description commonly taken to involve the use of computers in calculation, prediction, simulation, information storage and retrieval, etc. Computers also facilitate drug design through their impact on the technology of measurement automation.

In a situation where, for example, ten experimental options are available, CADD has a traditional role in selecting the five most likely to be informative. Recent advances in computer-driven technology have brought about a situation in which it is sometimes more effective simply to *do* all ten experiments.

Experience gained from setting up automation projects in drug design has shown the importance of identifying those tasks which are best done by machines and those which are best (or can only be) done by people.

The underlying theme of this presentation is that a tendency is emerging to use the computer as an automation tool in lead generation. This can allow more medicinal chemistry resource, backed by predictive CADD techniques, to be dedicated to the optimisation process, where it is generally in short supply.

The discussion, in three parts, will hopefully illustrate some ways in which we are currently exploiting the many CADD opportunities. The first section covers our perception of the importance of macromolecular structure information in the light of improved technology which has speeded up the process of data capture and interpretation. In the second section, the real objectives of lead optimisation are examined and changes in approach are exemplified. Finally, the impact of screening and related technologies is discussed.

11.2 The Role of Macromolecular Structure

The availability, mainly from X-ray crystallography, of high resolution three-dimensional (3D) structures of proteins is an attractive prospect for the drug researcher. If you have in front of you the molecular structure of the proposed target for drug action – what better starting point?

Much of the controversy which has surrounded this approach has arisen from overenthusiasm and consequent overexpectation. In real life it has proved itself to be an important part of our approach to molecular design which begs to be used whenever it is available. It is commonly proving its worth both in lead generation and in lead optimisation (Bugg et al. 1993; Navia and Murcko 1992).

In ligand design it offers both a source of inspiration and a vehicle for predictive modelling so that ideas can be "screened" for suitability. It should not be judged as an exclusive approach in which a single concentrated period of modelling, based on either an empty binding site or a site containing a natural substrate, is expected to be sufficient to take the medicinal chemist effortlessly through to a sturdy, robust development compound. The reality is very different. In practice, the approach is just one component method among many contributing extra insight to the discovery process.

Many early shortcomings have been overcome by advances in technology. The number of proteins for which structural data were potentially available was limited and the time taken to obtain the data did not fit the time scale of a normal research programme. Advances in molecular biology have transformed the situation as far as availability of protein is concerned. Opportunities also now exist for site-directed mutagenesis to gather information about the importance of interactions with specific residues.

The process of gathering and processing diffraction data has also improved as have, of course, the facilities for displaying the data. Early work produced physical models of great ingenuity. These even reflected, in some cases mechanically, changes in protein structure. The power of computational chemistry, molecular graphics and 3D viewing equipment have transformed this interface as well.

Thus, in the space of 20 years, the technique has become applicable to a wider range of targets and on a time scale which is useful to the medicinal chemist. Advances in all facets of the process of obtaining structures have been such that ligand-bound information can be provided to the chemist in something approaching real-time. The IC_{50} and the log P may arrive first, but not by much.

Membrane-bound proteins remain a problem, although work on porin channels, for example, has demonstrated what can be done through simulation of the membrane environment using detergents (Kreusch et al. 1991). The complementary information provided by nuclear magnetic resonance (NMR) is also important in this context.

Another criticism of the approach is that a crystal structure is "non-physiological" and therefore of limited relevance to the real-life, wet, mobile world of protein–ligand interactions. However, even in the crystalline state the protein is well endowed with associated water molecules. A practical confirmation of this is that crystallographers are often able to soak a crystal of one protein–ligand complex in a solution of another ligand and achieve ligand exchange. Thus, there is at least a basic level of normal solution chemistry going on. Furthermore, consistency with NMR data and success in prediction (see below) provide further reassurance.

The availability of a ligand-bound structure is a more solid foundation for ligand design than the native-enzyme structure, although the latter can lead to interesting speculations on allosteric sites. With a ligand-bound structure as a starting point, two options are available. One is to "throw away" the ligand(s) and work with the empty site (albeit already "educated" by the imprint of a ligand). The other is to retain the ligand(s) in the model and seek to improve its potency and/or specificity through modifications which exploit hitherto unused binding possibilities.

The latter, though more conservative, carries fewer risks associated with protein flexibility, and does allow modification which is much more radical than anything which could be achieved by quantitative structure–activity relationships (QSAR) or pharmacophore modelling. This is essentially lead optimisation or secondary lead generation. It is based on measured data and can be pursued one stage at a time through the following sequence:

1. Inspection of ligand-bound structure
2. Identification of new binding opportunity

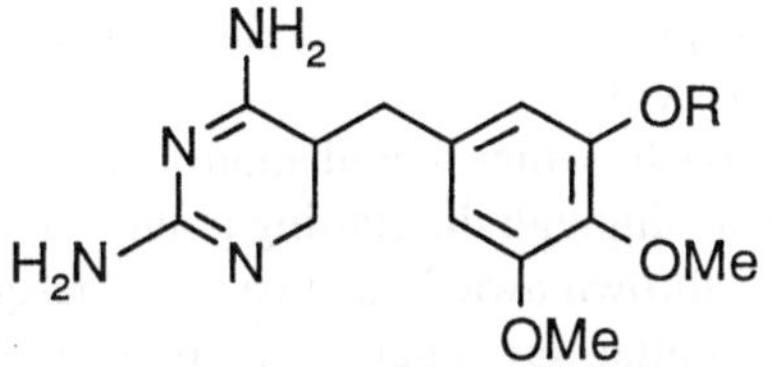

R	K_D (relative to II)
(I) -(CH$_2$)$_5$COOH	0.063
(II) -Me	1.0
(III) -(CH$_2$)$_3$COOH	0.15

Fig. 1. Inhibitors of dihydrofolate reductase *Escherichia coli*

3. Modelling proposed compound
4. Synthesis
5. Biochemical evaluation
6. Co-crystallisation and structure determination of new complex

This process was exemplified at Wellcome (Kuyper et al. 1985). Compound I (Fig. 1) had been designed using 3D molecular models of the complex between dihydrofolate reductase (DHFR; *Escherichia coli*) and methotrexate. Higher affinity was sought by introducing a carboxyl group on the end of a methylene chain. This was intended to interact with an arginine residue (Arg 57).

The compound showed a more than 50-fold increase in affinity for the enzyme compared with trimethoprim (II) (Fig. 1). Analogues having shorter chain length, e.g. (III; Fig. 1), showed lower affinity as expected from the model. The X-ray crystallographic studies which were done on trimethoprim, I and III confirmed the expected binding mode and were consistent both with the expectation from modelling and with the measured affinity constants.

This early exercise ably demonstrates that when working in this lead optimisation mode with a known site and using measured data, the information from the crystalline state *does* provide a useful basis for real chemical improvement, and that its capacity to mislead has sometimes been overstated. With today's improved data collection and modelling facilities, this process can be undertaken on a routine basis and on a much shorter time scale.

The other approach, which could more correctly be seen as lead generation, is when, although the starting point is a ligand-bound structure, the ligand is "thrown away", and de novo design is attempted for the empty site. Potential new ligands can be generated intuitively, or though a number of possible CADD methods (Kollman 1994, Goodford 1985, Glen 1994).

Progress in this field is rapid and in future the prospect of more quantitative modelling can be entertained. At present a list of putative structures can be generated and modelled to assess their fit. To be able to confidently rank them in order of predicted binding constants and synthetic feasibility is a sought-after future goal.

Beyond that, it is essential to focus on the need for CADD techniques to contribute to the discovery of new drugs (or at least robust development candidates). Producing exquisite enzyme inhibitors is not enough.

The contribution (albeit indirect) of structural information to the other processes of optimisation – i.e. those concerned with drug access to target – will be discussed in the next section.

11.3 Re-Evaluation of Lead Optimisation

The process of getting from a lead molecule to a candidate for development involves more than optimising potency at the enzyme (or receptor) target. The chosen molecule must also have the right properties to allow it to gain access to the target at an effective concentration and on an appropriate time scale.

One approach to lead optimisation is to concentrate exclusively on increasing target potency, then select the most potent compound and hope to sort out absorption, distribution, metabolism and elimination (ADME) problems later. This sequential approach is wasteful in terms of resources. Furthermore, it can delay the identification of insuperable ADME problems.

An alternative approach is to build up from the start a broader appreciation of the multiple criteria which have ultimately to be met. The real objective of lead optimisation is the fulfilment of all these criteria with a single compound. If this is not possible, i.e. if criteria for

target activity and access to target are incompatible, this must be identified as soon as possible.

The whole emphasis should be on defining levels of acceptable
performance rather than identifying properties associated with peak
performance in any one of the relevant assays. Development candidates
are rarely the most potent enzyme inhibitors.

Defining acceptable performance is difficult; however, any attempt,
based on experience, is better than simply pursuing every single aspect
of the biological profile to its optimum. It is necessary to make qualitative judgements on quantitative data when optimising a lead. Data
cannot normally be "traded" in a quantitative manner.

Criteria for action at the target are commonly expressed in structural
terms, i.e. a more or less formal expression of a pharmacophore. Criteria
for access, with the exception of those dependent on metabolic processes, are often expressed in terms of physicochemical attributes, such
as solubility, partition, ionisation and molecular size.

In establishing these physicochemical criteria, biological data are
required. Sometimes, information about ADME processes is available
for the series of compounds being optimised. This information may take
the form of a direct measure of concentration, e.g. cell uptake or al-
bumin binding. Alternatively, the ADME component may be inferable
from measures of effect in assay systems of different complexity. This
sort of data is useful in identifying a problem and in testing hypotheses
for solving problems.

However, in order fully to explore the limits of acceptable physicochemical properties associated with the required in vivo profile – and to
do so early enough – a broader data base than the immediate series of
interest should be referred to.

The benefits of using information on a diverse set of chemicals to
suggest regions of acceptable physicochemical space generally have to be
bought by accepting data of lower precision. Provided the interpretation is
consistent with the nature of the data, this does not have to be a problem.

Thus, an approach to lead optimisation which has been successfully
adopted is to seek precedent for satisfactory in vivo performance from
data bases or literature reports. Searches based on substructure or on
physicochemical properties can allow a picture to be built up of the
likelihood of a particular group of lead molecules being able to fulfil
their potential.

If there is no precedent in terms of the properties searched, then a judgement can be made as to the extent of molecular modification needed to allow activity in cells. The likelihood of this modification being compatible with retention of activity at the target can be assessed using a QSAR, a pharmacophore model, or, best of all, a comprehensive appreciation of inhibitor binding based on measured ligand-bound structural data.

This sort of information base can allow required chemical modification to be made in radical ways, using, if necessary, binding opportunities involving residues not exploited hitherto. It is in this way that the measured structural information for ligand–target interaction can be a most positive benefit in the process of producing a robust development candidate.

11.4 CADD Applied to a Series of 5HT$_{1D}$ Agonists

A real-life example of this type of approach (Hill et al. 1994) was encountered in the search for new molecules for the treatment of migraine (Martin et al. 1994,). In this case no macromolecular structural data were available, although there was a well developed SAR and a pharmacophore model to work with (Glen et al. 1994).

A number of selective and potent agonists of 5HT$_{1D}$ receptors had been identified in in vitro assays. They had been found to lack in vivo activity and, following measurement of plasma concentrations, this was attributed to low oral bioavailability resulting from poor absorption from the gastrointestinal (GI) tract.

Literature reports (Lien 1974) of QSAR studies on GI absorption indicate optima corresponding to log D values considerably in excess of those for the 5HT$_{1D}$ agonists, which were all hydrophilic, having log Ds around or less than zero. The problem was therefore approached by asking the simple question – are there any orally absorbed drugs with this physicochemical profile?

For a set of compounds known to be orally administered (Sharman 1969), log D and molar refraction (MR) values were calculated (Pomona 1989). These properties were chosen to describe attributes most likely to be of importance for the two most commonly encountered nonspecific mechanisms of absorption. These are the transcellular

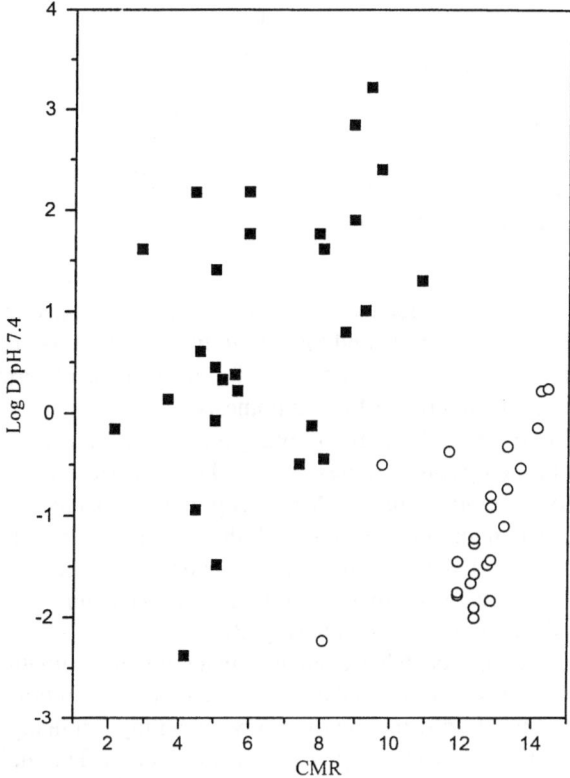

Fig. 2. Disposition of compound sets in log D/CMR descriptor space. *Squares,* literature compounds; *circles,* early 5HT$_{1D}$ agonists

route, involving diffusion across a membrane, and the paracellular route, involving the use of aqueous pores (Taylor et al. 1989).

Log D and MR were then also calculated for the set of 5HT$_{1D}$ agonists. The situation which was found is summarised diagrammatically in Fig. 2. The hypothesis which was proposed on the basis of these observations was that the agonists were too hydrophilic to avail themselves of the transcellular (diffusion) route and yet just too large to gain access via aqueous pores (the paracellular route).

There was one data point which was poorly fit. It represented a compound whose MR was just low enough to enter the "allowed" zone,

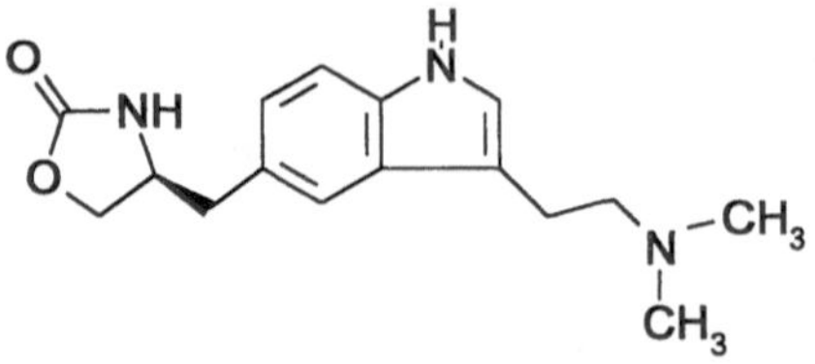

Fig. 3. The structure of 311C90

but yet was one of the nonabsorbed agonists. However, it was the most hydrophilic of the agonists and the possibility that the expected greater degree of hydration was responsible for its lack of absorption was entertained in the interests of the hypothesis.

Of the two possibilities for moving into the desired region of log D/MR data, the option of increasing log D was ruled out on practical grounds. An attempt to reduce MR was preferred. A reconciliation of this physicochemical information with the existing SAR and the pharmacophore model derived from it was sought. A number of orally absorbed agonists were then identified and the compound 311C90 chosen as development candidate (Fig. 3).

This general approach has been used in a number of situations and is capable of extension. Improved data bases and better searching software allow us to unlock information on in vivo handling originally obtained for compounds of a wide range of therapeutic class. This information can help us to decide whether activity and access properties are likely to be reconcilable and can suggest part of how to achieve the reconciliation.

The preceding section illustrates a process of identifying criteria for components of an acceptable drug profile and seeking to reconcile them. The better the target activity component of the overall profile is defined, the easier it is for the medicinal chemist to devise ways of meeting physicochemical criteria for "drug access" while retaining the essential activity. Where the target activity criteria are illuminated by structural information, options for making radical chemical modifications to accommodate substantial physicochemical changes demanded by a physicochemical model for drug access become clearer.

Thus, structural information can contribute importantly to the optimisation of leads. The ability to return to a visual representation of a ligand in a protein binding site is one way of helping to modify chemistry to achieve in vivo activity.

Another way is to make use of technological advances in screening. If it becomes clear that a lead cannot reasonably be optimised, then screening can allow us to throw that lead away and start using another one. This concept – the disposable lead – is discussed in the next section.

11.5 The Return of Screening

The re-emergence of screening as a drug discovery technique has arisen from changes in both technology and culture. Although apparently at the opposite end of the scale from structure-based design, it provides an effective complement to the "rational" approach and involves not only its own elegant technology, but also its own intellectual challenges.

Closely associated with screening programmes are the technologies of combinational libraries and array synthesis which have the potential to revolutionise lead generation (De Witt et al. 1993; Gallop et al. 1994; Gordon et al. 1994). Taken together with the structural approach (where applicable) we are progressing towards a position in which, if the first lead proves hard to optimise (in the full sense of the word), one can go back and try another one. This should improve the success rate in terms of providing robust development candidates. Fruitless effort spent in trying to impose acceptable ADME properties on a "difficult" lead molecule can be better spent elsewhere.

The benefits of a screening approach are that it offers the possibility of novel leads, i.e. ones which do not look like the natural substrate, and multiple leads, with the advantages mentioned above. It also carries the advantage of economy in "recycling" existing chemistry.

11.6 Using Existing Chemistry

The benefits of using existing chemistry at the lead generation stage are just as pertinent to situations in which there is some structural information as a starting point. Hypotheses and speculations can be tested out using existing chemistry free from the constraints of competition for a generally overstretched synthetic resource.

Molecules originally synthesised in pursuit of one specific target are potential probes of a range of biological systems. They represent a limited number of well-characterised functionalities topologically disposed in a way which is known. Improvements in methods for conformational prediction mean that these existing molecules can be used as 3D probes to extract criteria for binding through a screening exercise.

To use them only once is therefore a wasted opportunity. Thus, if potential therapeutic targets can be explored with the resources of just an assay, and a library of existing compounds, then a greater number can be assessed for suitability. This should allow better decisions to be made about which targets to pursue, since a proposed target for which a chemical lead has already been found will argue its case more strongly than others.

In order to exploit properly the resource of a library of existing compounds, the need for an automated compound handling and prescreen formulation facility was recognised. The driving force for the introduction of this facility, to be known as HAYSTACK, was the increasing use, with some success, of the in-house compound library in the search for leads. Prior to HAYSTACK, two approaches, "intelligent" screening and "educated" screening, had been used, both relying heavily on CADD resources.

For "intelligent" screening, subsets of compounds are selected for assay so as to give maximum information. The objective is to achieve a spread in structural and physicochemical properties (Johnson et al. 1989). It is assumed that a conscious attempt to achieve diversity, in a set of compounds based on experience, is better than relying on purely random selection from a compound library.

To meet this requirement a set of compounds was drawn up at Wellcome to represent the chemical diversity of its store. A prerequisite was adequate availability, and, indeed, such practical prerequisites are often the first stage in set selection. The second stage was simply to interview the staff of the Medicinal Chemistry Department to ask them to select a set of compounds which they believed represented their work over the years. Where possible, some hierarchy was introduced, so that there was a size option in set selection. This set proved useful, was popular, and did much to lay the foundations for later developments.

The need for a more objective approach was recognised, as was the need for greater flexibility. If set selection methods could be improved and linked to a store of readily accessible compounds of assured availability, ad hoc sets of any size could be supplied in addition to standard sets.

Standard sets have two main advantages. They are convenient and hence popular with screeners. Secondly, by putting the same compounds through a large number of screens, activity–activity relationships can be built up. On the other hand, ad hoc sets can be tailored to the requirements of any particular screen and help avoid the practical problem of depleting compound libraries of key compounds.

Consequently, a mixed approach has been adopted. Standard sets, based on compounds which are relatively abundant, are being made available, while the facilities for choosing specific ad hoc sets is also in place.

The main reason why the principles of set selection familiar in lead optimisation (Martin and Panas 1979; Craig 1971) could not be more widely applied was a lack of opportunity for molecular description for diverse sets of compounds. Clearly, substituent constants were of no use.

The possibility of calculating log P and MR for diverse molecular types together with the availability of other readily accessible descriptors has opened up the field further and selection based on these properties has led to the establishment of useful compound sets.

With the advent of readily available calculated 3D structures (CONCORD) and subsequent large scale property calculation (Glen and Rose 1987), it seemed that the way was clear to set up huge data matrices of steric, electronic and hydrophobic properties on which to base compound selection. A comparison has been made using Procrustes Analysis (Rose et al. 1994) of the performance of three types of descriptor sets in the characterisation of compound sets. The first type is based on readily accessible 2D physicochemical properties such as CLOGP, CMR (DAYLIGHT) and Andrew's Binding Energy (Andrews et al. 1984). The second type is based on properties calculated (PROFILES) from overlaid 3D molecular structures. The third type is based on connectivity indices (Kier and Hall 1976).

For the 14 sets investigated, three distinct clusters are visible. The authors conclude that all the sets give logical groupings but that each is telling us something different about the molecules from which sets are to be selected.

In practice, criteria such as availability, abundance and suitability for assay conditions are often used as a primary filter, prior to selection on the basis of calculated properties. In some cases, for example, assay conditions demand limits of log P or superior aqueous solubility.

After having filtered the population by practical criteria, molecular descriptors are calculated and a data matrix drawn up. The final set may be selected either by methods based on samples of the population (e.g. cluster analysis) or sampling the space (e.g. sphere exclusion). It must be borne in mind that the starting population available in company compound stores reflects the areas which have been worked on and is not necessarily a comprehensive and balanced source of probes.

The success of lead generation based on the frequent use of low/medium throughput assays will depend on the quality of this set selection process and remains an important area for CADD in the highly automated environment of screening.

The process of "intelligent" set selection can be applied in cases where there is no suggestion of a chemical starting point. However, it is not uncommon to meet a situation in which some broad chemical pointers are available, in which case a process of "educated screening" can be undertaken.

The term "educated screening" is here used to describe set selection which is based on similarity searches. The searches may be based on analogy to existing "actives" reported in the literature, natural substrates, a pharmacophoric model derived from a combination of known "actives" or a complement to a binding site. The facilities for 2D or 3D searches have transformed this process over the past 10 years.

It became clear that, with improved search and modelling facilities, and assays capable of higher throughput, an efficient procedure for the retrieval, weighing and formulation of existing compounds was urgently required. Such a procedure would be essential if low throughput screening were to be properly exploited and if high throughput screening were even to be contemplated.

"High throughput screening" has now been made possible in this laboratory by the installation of an automated compound handling facility HAYSTACK. This allows retrieval, weighing, dissolving and aliquotting to be performed by robot. Frozen liquid samples are dispensed to biologists in standard 96-well plate format. A manual weighing station deals with those compounds which cannot be handled by the

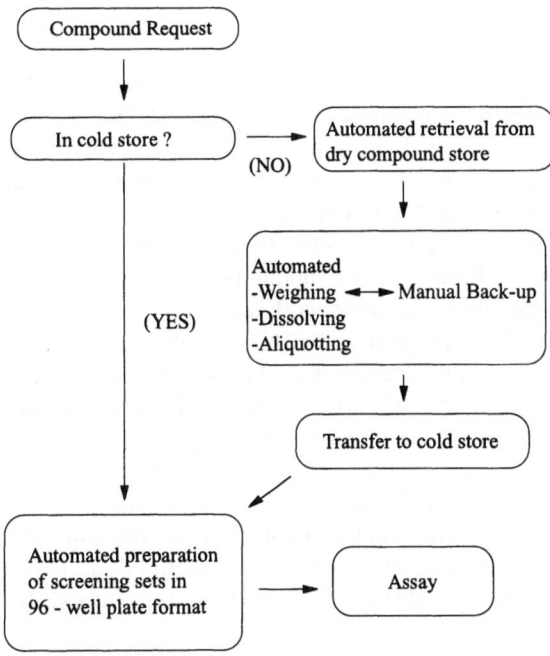

Fig. 4. The HAYSTACK process

automatic weighing system. The key features of the system are summarised in Fig. 4.

Apart from the fact that this is a sine qua non for high throughput
screening, it greatly improves the efficiency of low throughput work
and essentially unlocks the investigative power of a library of existing
chemical probes.

The HAYSTACK system has provided not only a key facility, but
also a focus for the growing range of activities associated with screening. In addition to the availability of technology for selecting, retrieving
and formulating existing compounds, automation opportunities now
exist for synthesis itself. Medicinal chemistry and CADD are respond-
ing rapidly to these new opportunities.

11.7 Conclusions

1. Advances in technology at all stages have increased the power of CADD based on macromolecular structure. The approach is more than an inspirational aid to primary lead generation, and its role in lead optimisation, based on real measured data, is proving to be as least as important if not more so.
2. Lead optimisation is too often focused exclusively on increasing potency. Multiple criteria should be addressed early, if only to make a case for terminating work on a series.
3. By providing comprehensive insight into the environment of a bound ligand, the structure-based approach can play an important role in lead optimisation. Molecular modification required to clear hurdles between initial lead and development candidate can be modelled to assess the chances of retaining the core activity. Priorities for synthesis can be more easily identified.
4. Advances in automation technology have brought screening approaches back to prominence. CADD methods are essential in exploiting screening through set selection and effective lead follow-up.
5. Screening methods are complementary to the structure-based approaches, underlining the view that modern approaches to drug design must be flexible, broad and based soundly in interdisciplinary cooperation.
6. Improved technology both for screening and structure determination allows the resources of synthetic medicinal chemistry backed by CADD to be concentrated on the process of producing development candidates from leads.

Acknowledgements. The author is grateful to the following colleagues for their contributions, advice and support: Alan Hill, Anne Hersey, Bobby Glen, John Champness, Brian Hudson, Ken Powell, Rob Lifely, Lee Kuyper, Sally Rose, Alan Robertson, Graeme Martin, Jeremy Stables, and Willy Harrison.

References

Andrews PR, Craik DK, Martin JL (1984) Functional group contributions to drug-receptor interactions. J Med Chem 27:1648–1657

Bugg CE, Carson WM, Montgomery JA (1993) Drugs by design. Sci Am 60–66

Craig PN (1971) Interdependence between physical parameters and selection of substituent groups for correlation studies. J Med Chem 14:680–684

DAYLIGHT chemical information systems, 3951 Claremont Street, Irvine, California, 92714, USA

De Witt SH, Kiely JS, Stankovic CJ, Schroeder MC, Cody DMR, Pavia MR (1993) Diversomers: an approach to non peptide, non oligomeric chemical diversity. Proc Natl Acad Sci USA 90:6909–6913

Gallop MA, Barrett RW, Dower WJ, Fodor SPA, Gordon EM (1994) Applications of combinatorial technologies to drug discovery. 1. Background and peptide combinatorial libraries. J Med Chem 37:1233–1248

Glen RC (1994) Genetic algorithms. J Comput Aided Mol Design (in press)

Glen RC, Rose VS (1987) Computer program suite for the calculation, storage and manipulation of molecular property and activity descriptors. J Mol Graph 5:79–86

Glen RC, Hill AP, Martin GR, Robertson AD (1994) Molecular design of 5-HT_{1D} agonists for the acute treatment of migraine. Headache 34:307

Goodford PJ (1985) A computational procedure for determining energetically favourable binding sites on biologically important macromolecules. J Med Chem 28:849–857

Gordon EM, Barratt RW, Dower WJ, Fodor SPA, Gallop MA (1994) Application for combinatorial technologies to drug discovery. 2. Combinatorial organic synthesis, library screening strategies and future directions. J Med Chem 37:1387–1401

Hill AP, Hyde RM, Robertson AD, Woollard PM, Glen RC, Martin GR (1994) Oral delivery of 5-HT_{1D} receptor agonists: towards the discovery of 311C90, a novel anti-migraine agent. Headache 34:308

Johnson M, Lajiness M, Maggoria G (1989) Molecular similarity: a basis for designing drug screening programmes. In: Fauchère JL (ed) QSAR in drug design. Liss, New York, pp 167–172

Kier LB, Hall LH (1976) Molecular connectivity in chemistry and drug research. Academic, New York (Medicinal chemistry monographics, vol 14)

Kollman PA (1994) Theory of macromolecule-ligand interactions. Curr Opin Struct Biol 4:240–245

Kreusch A, Weiss MS, Welte W, Weckeiser J, Schulz GE (1991) Crystals of an integral membrane protein diffracting to 1.8 Å resolution. J Mol Biol 217:9–10

Kuyper LF, Roth B, Baccanari DP, Ferone R, Beddell CR, Champness JN, Stammers DK, Dann JG, Norrington FE, Baker DJ (1985) Receptor-based design of dihydrofolate reductase inhibitors: comparison of crystallographically determined enzyme binding with enzyme affinity in a series of carboxy-substituted trimethoprim analogues. J Med Chem 28:303–311

Lien E (1974) Relationship between chemical structure and drug absorption, distribution and excretion. In: Maas J (ed) Medical chemistry, proceedings of an international symposium. Elsevier, Amsterdam, pp 319–342

Martin YC, Panas HN (1979) Mathematical considerations in series design. J Med Chem 22:784–791

Martin GR, Robertson AD, MacLennon SJ, Prentice DJ, Honey A, Barrett VJ (1994) A novel, selective 5-HT$_{1D}$ receptor partial agonist for the acute treatment of migraine. Headache 34:305

Navia MA, Murcko MA (1992) Use of structural information in drug design. Curr Opin Struct Biol 2:202–210

Pomona (1989) Physicochemical database and medchem software, Version 3.54, Daylight Chemical Information Systems, Inc, Claremont, California, USA

Rose VS, Rahr E, Hudson BD (1994) The use of procrustes analysis to compare different property sets for the characterisation of diverse compounds. Quant Struct Act Relat 13:152–158

Sharman DF (1969) Drugs and other pharmacologically active compounds. In: Dawson RMC, Elliot DC, Elliot WH, Jones KM (eds) Data for biochemical research. Oxford University Press, Oxford, pp 405–421

Taylor DC, Lynch J, Leahy DE (1989) Modes for intestinal permeability to drugs. In: Hardy JG, Davis SS, Wilson CG (eds) Drug delivery to the gastrointestinal tract. Ellis-Horwood, Chichester, pp 134–135

12 Theoretical Chemistry as Part of the Interdisciplinary Approach to Rational Drug Design

E. Eckle and N. Heinrich

12.1 Introduction

The early days of applied computational chemistry were characterized by two extreme attitudes, best represented by E. Clementi's enthusiastic statement in 1973 "We can calculate everything!" and C. Coulson's reply "Give us insight not numbers!" A further decade was needed to convincingly demonstrate that computational methods can contribute significantly to real problems in organic chemistry. Theoretical methods now provide an invaluable tool to rationalize many aspects of chemical reactivity. As a consequence of the enormous developments in hardware and software, quantum mechanical calculations with a fairly high degree of sophistication can now be applied relatively easy even to series of medium-sized organic molecules such as steroids. For drug design the developments in computer graphics certainly played the decisive role to visualize molecular properties and to communicate theoretical results to the experimentalists. The potential of molecular modeling techniques as an inspiring tool to create new and unconventional ideas has now widely been recognized. The contributions from computational chemistry can guide a drug finding process into areas of new (lead) structures. However, no one discipline ever produces a drug, so computational chemists cannot either. Among other areas such as synthetic chemistry, pharmacology, and biology, computational chemistry is one part of the interdisciplinary approach in the early stages of developing a drug.

During the last decade, developments in theoretical chemistry have been challenged and inspired by the remarkable progress made in certain experimental disciplines: By means of molecular biology techniques, large amounts of virtually any protein can be obtained. Improved purification methods and the progress in structure elucidation techniques, such as protein nuclear magnetic resonance (NMR) methods and protein crystallography make it possible to get detailed information about the three-dimensional (3D) structures of potential targets for pharmaceutical research. The computational aspect of properly tackling large biomolecules has been facilitated by developments of improved force fields, protein dynamics and free-energy perturbation methods for an adequate description of protein–ligand interactions (van Gunsteren and Berendsen 1990; Lybrand 1990). Docking procedures, 3D database searching techniques and de novo methods have been described to design new ligands into well-characterized binding sites (Verlinde and Hol 1994; Martin 1992).

For lead discovery, structure-based ligand design has proven to be one of the most powerful and promising approaches to a rational design. Some compounds derived from these techniques have already been successfully tested in clinical trials. As long as chemical synthesis is still one of the rate-determining steps in drug design programs, the target structure-based approach is certainly the most rigorous answer to the conventional, mainly chemically driven, strategy. In order to discover a tightly binding ligand in a shorter period of time, a directed synthesis of well-designed compounds is being used to replace the enormous efforts required to identify the essential functionality by chemical means. A detailed knowledge about the binding cavity of the protein may, in addition, provide hints as to which positions are best suited for modifications in order to tune the physicochemical properties in the subsequent optimization process. Compared to classical strategies, however, structure-based ligand design will have to prove its advantage, in particular in terms of time reduction: The rate-determining step of this approach has now been shifted towards operations such as protein overexpression, purification, crystallization, and structure elucidation, which may also be time consuming.

A completely different approach may be seen in the recent developments in combinatorial chemistry linked to high-capacity screening methods (Alper 1994; Gallop et al. 1994). Here, the time for synthesis and biological testing is no longer expected to determine the rate of progress. Combinatorial libraries can be produced in-house or can be obtained from external sources, such as university groups or commercial vendors. Before testing, much effort has to be put into the optimization and automatization of binding and/or functional assays for the particular conditions of high throughput. The understanding of drug interactions on the molecular level, however, does no longer play the decisive role. Following this approach, initial lead structures are found rather than designed. According to expectations, this might be achieved within a distinctly reduced period of time compared to other strategies. Once a lead has been discovered, the subsequent chemical optimization, however, cannot benefit from any information on the functionality necessary to retain the biological activity, as is the case for structure-based ligand design. The essential groups have to be identified by systematic chemical variation before certain physicochemical properties can be adjusted. Statistical methods (Franke 1984) come into play when huge amounts of data from the high-capacity screening have to be managed and evaluated in order to find relationships between chemical structures and their activity profiles. Another area to support this approach by theoretical methods may be seen in the field of molecular similarity (Johnson and Maggiora 1990) applied for set selection in mass screening programs.

The two approaches briefly outlined above are inherently different. They provide the two extremes within the spectrum of strategies whose success is measured in terms of potential candidates for development. At present, no one can truly say which concept will be the more successful in the future. Theory groups have to cope with these diverging trends.

Depending on whether the 3D structure of the biological target can be derived, "direct" or "indirect" modeling strategies have to be applied. How do models for structure–activity relationships resulting from these two strategies compare with regard to their predictive power? Apart from the degree of sophistication in the description of structures, the quality of the relationships, i.e., the predictive power of the model, depends to a major extent upon the uncertainties in the biological data.

In general, we have to deal with models not only for structure but also for biological activity. "A model ... is a deliberately false, but, when taken in a given context, nevertheless expedient assumption" (according to Primas and Müller-Herold 1984). Each component of what we summarize under the term "biological activity" is determined in experiments under more or less artificial but well-defined model conditions. Correspondingly, representations of "chemical structure" may range from a simple topological description to structural models derived from quantum mechanical calculations. Each model has its own capabilities and limitations to approximate certain aspects of reality. Hence, structure–activity relationships provide (super) models with all inherent uncertainties and limits being retained from their theoretical and experimental components.

The most predictive models can be derived if the 3D structure of a target–ligand complex has been determined experimentally. Here, detailed information on the architecture of the active site as well as the binding mode and the bioactive conformation of the ligand is provided. Even in this ideal situation, however, attempts to calculate accurately or even to predict in qualitative terms the free energies of binding may fail due to limitations in our treatment of intermolecular interactions (van Gunsteren and Mark 1992).

If the 3D structure of the target has to be derived from homologous experimental structures, the resulting model may have limited validity due to the modifications to the parent structure, such as insertions, deletions, and replacements of amino acid residues. If these operations had to be made close to the ligand-binding region, they may corrupt any correlation between theoretically derived features and measured affinity data in the "semi-direct" approach.

However, if there is no information on the biological target structure, the model will be based on comparisons of active and inactive compounds to sort out the key features responsible for activity. This "indirect" approach exclusively depends on a reliable theoretical representation of the ligand structures, i.e., their conformational flexibility and their electronic properties, as these factors determine the way the structures are to be superimposed. In order to assure the most efficient use of theoretical resources, a detailed analysis of the available biological data has to precede any modeling activity in order to select an appropriate theoretical method. For example, typically "weak" biological data such as scoring data from greenhouse experiments in herbicide research, do not need to be compared with results from highly sophisticated ab initio calculations as the biological response can, at most, be classified into the three categories "active," "medium-active," and "inactive." A similar but correct grouping of theoretically derived features ought to be sufficient to discover the trends within the frame of a relatively "weak" qualitative model. Nevertheless, to recognize the trends at relatively early stages of the drug finding process often has a substantial impact on the progress although an exhaustive understanding of underlying principles has not been achieved.

How accurate does a model need to be to provide a sound basis for suggestions of ligand structure modifications? When should a theory group start to support a project? These two questions are, indeed, closely related to each other: Drug research programs in industry have to demonstrate progress on a regular basis. A compromise between the predictive power of a model and the effort in time to generate the model often has to be made. Strategic experiments to improve the model, e.g., site-directed mutagenesis or the synthesis of complicated model compounds, have to fit reasonably well into the time schedule of the drug finding program. Moreover, their expense in costs and time has to be justified by the expected gain of useful knowledge to further validate a model. An optimum of mutual benefit needs a steady exchange of even preliminary results and can be achieved only if the process of model development and the experimental activities proceed in parallel. Otherwise, the model may give, at best, a nice but a posteriori explanation.

12.2 Ligand-Based Models

Unfortunately, in most of the drug finding projects no information is available about the 3D structure of the macromolecular target enzyme or receptor. The development of a model for structure–activity correlations, therefore, has to be performed "indirectly." As already mentioned

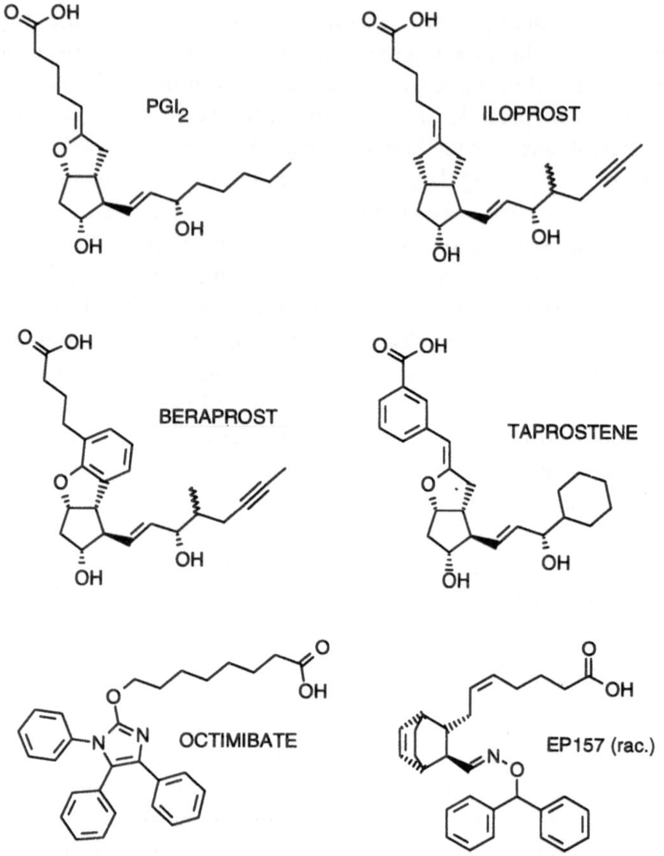

Fig. 1. Chemical structures of prostacyclin (PGI₂) and selected prostanoid and nonprostanoid PGI₂ receptor agonists

above, this approach is based purely on comparisons of stereochemical and physicochemical properties of a set of well-selected ligand structures.

The modeling of non-prostanoid prostacyclin (PGI₂) agonists is an appropriate example to further explain this approach. PGI₂, a product of the arachidonic acid cascade, is known to be a very potent vasodilator and inhibitor of platelet aggregation, exerting its action via specific receptors in the platelets and vessels. Due to its chemical instability, however, its use as a drug for antithrombotic therapy is rather limited. At Schering, a drug finding project was started to develop a chemically stable analog with a similar pharmacological profile, thereby being orally available. Iloprost turned out to be the most promising compound in several pharmacological tests (Gryglewski and Stock 1987) and has been passed on for clinical development (Skuballa and Schäfer 1989). At the same time, other companies came up, for example, with structures such as taprostene and beraprost (see Fig. 1) as potential candidates for development (Skuballa and Schäfer 1989). All of these molecules, however, are closely related to the complicated architecture of the natural compound PGI$_2$. Rather lengthy synthetic pathways had to be followed to introduce the correct stereochemistry of the chiral centers as their specific absolute configurations were known to essential for high biological activity. On the other hand, octimibate was reported to be an albeit weaker PGI$_2$ mimic acting through the same type of receptors (Seiler et al. 1990; Merritt et al. 1991). Compared to the prostanoid-type structures this chemically rather simple species was seen as a promising starting point for the design of a new nonprostanoid PGI$_2$ agonistic lead structure.

First of all, a 3D pharmacophore model for prostanoid and nonprostanoid structures had to be developed in order to learn more about the spatial and functional correspondence of the two classes of structures. Therefore, for each of the ligands a well-founded approximation to the receptor-bound conformation had to be made following the principles of the active analogue approach (Marshall et al. 1979).

Structure–activity relationships of prostanoid-type structures lead to a reasonable guess to the topological pharmacophore pattern, as indicated in Fig. 2. The acidic function in position 1, the hydroxy group in position 15, and two lipophilic moieties specified by the bicyclic head group and the area beyond position 15 have been assigned to be the key features.

Corresponding to Fig. 2, the complementary areas at the hypothetical receptor binding site are designated as P1/P2 and L1/L2/L3 for polar and hydrophobic regions, respectively. Due to a lack of structures closely related to octimibate, its functionality essential for activity could, however, not be assigned unequivocally. The same was true for the compound EP 157 which had been shown to have PGI$_2$-like activity

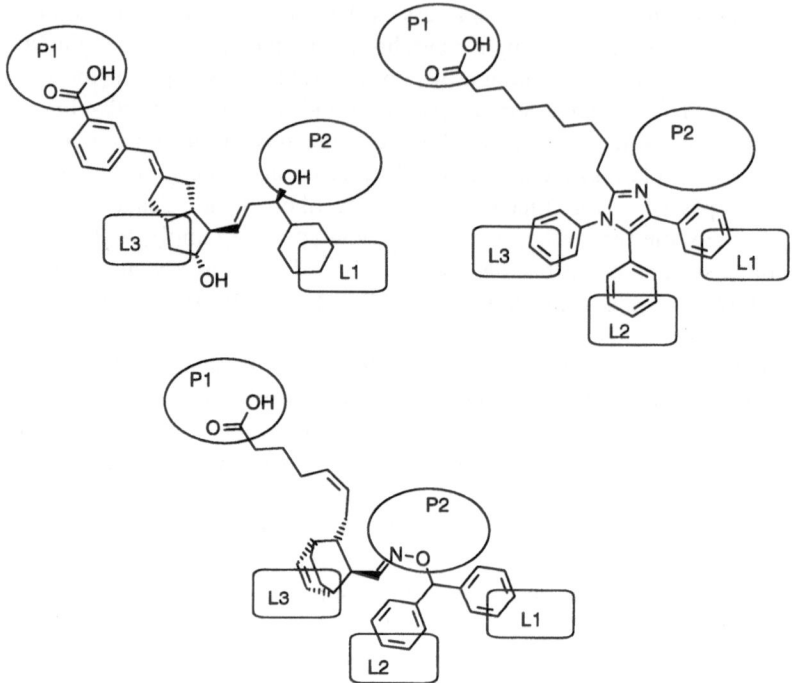

Fig. 2. Topological pharmacophoric pattern for PGI$_2$ receptor agonists. Polar and hydrophobic regions at the hypothetical receptor binding site are designated as *P1/P2* and *L1/L2/L3*, respectively

mediated through the prostacyclin receptor (Armstrong et al. 1986). Apparently, the structure of EP 157 may be seen as a hybrid, combining features from both types of structures, the prostanoids and octimibate, respectively. Therefore, the assignment of pharmacophore groups according to Fig. 2 provides a reasonable working hypothesis.

Systematic conformational analysis has been performed for the entire set of structures. Only those low-energy conformations were further considered which fit properly into a common 3D arrangement of the essential pharmacophore elements. Distance geometry methods (Sheridan et al. 1986) have been particularly helpful to get first proposals for common superpositions. In this approach, however, the energetics are

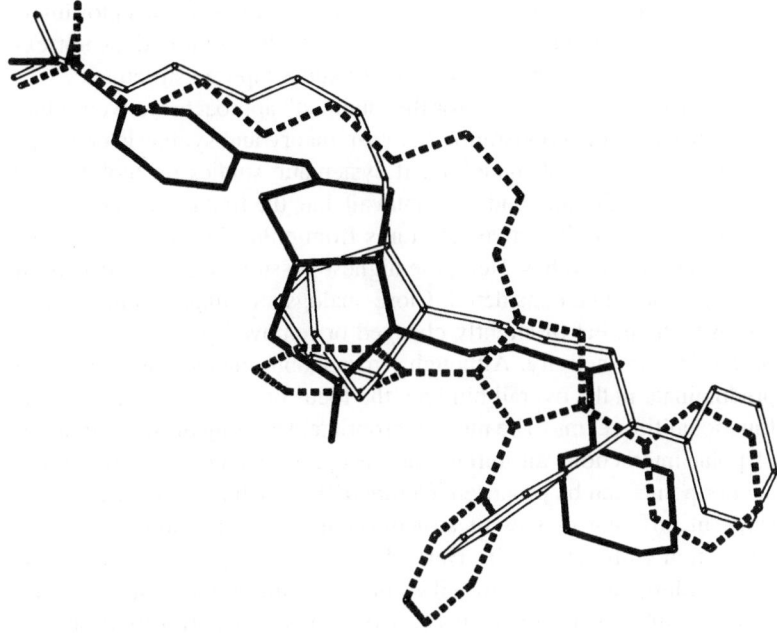

Fig. 3. Superposition of the 3D structures of taprostene (*thick bonds*), EP157 (*open bonds*) and octimibate (*broken bonds*) according to one plausible 3D pharmacophore model

not explicitly considered. An iterative procedure of geometrical fit, subsequent geometry relaxation, and comparison with the global conformational minimum structures had to be performed to discriminate "high-energy" from "low-energy" superpositions. This converged to a set of four superpositions. Here, the relative energies needed to adopt the fitting conformations turned out to be within a reasonable range of 3 kcal/mol above the global minima. One of the possible "low-energy" superpositions of selected structures is shown in Fig. 3. Each of the four superpositions can be interpreted in terms of a model of a putative spatial pattern of ligands bound to the receptor.

Based on this model, we succeeded in designing a new, chemically simple, and patent-free ligand which is at least as potent as the parent octimibate structure. In this species, the key features for receptor inter-action were retained. Furthermore, its molecular volume does not ex-ceed the "active" volume defined by the superimposed structures.

As the example is typical for the "indirect" approach, we would like to discuss how the cooperation between theory and synthesis develops during the process of modeling. If systematic studies on biologically related series of compounds are notavailable, the first guess is based on a limited set of only active structures from either in-house sources or from literature searches. Here, the highest possible degree of structural diversity should be considered. Close analogs are suggested in order to discover the effect of slightly changed or removed polar functionality on the in vitro activity. Although hydrophobic interactions may even predominate in the overall binding, the focus on modifications of polar functionalities seems to be more appropriate: Utilizing the directionality of polar interactions an initial pharmacophore pattern of well-defined anchor points can be proposed. Comparable structure–activity relation-ships in different classes of structures may reflect a similar kind of interaction with the protein. Here, the basic assumption is that a com-mon binding mode exists for all compounds under study. Based on the putative functional equivalences, a preliminary superposition can be derived. The multiplicity of superposition modes is an inherent problem when models of highly flexible molecules are to be developed. The only way to overcome this problem is to restrict the conformational space by the synthesis of rigid analog structures. If the activity of these structures is retained, they might be able to discriminate against alternative super-positions. Interesting skeletons for further chemical investigation could already appear at this stage. Based on the refined model further sugges-tions are then made to explore regions in space which exceed the common "active" volume defined by the training set of basic structures. Methyl or phenyl groups might serve as appropriate probes for steric mapping. A loss of activity due to a well-defined structural modification provides invaluable information on restrictions in the putative target site. A model can only display its predictive power in more than quali-tative terms, however, if the whole range of activities is sufficiently well covered by structures. The discovery of additional "active" areas in space may provide the desired basis to turn from closer analogs to suggestions with a higher degree of structural diversity. However, with increasing knowledge resulting in more elaborate suggestions for tightly binding ligands, the need to focus the synthesis program are becoming more important. Therefore, to attract the attention of the synthetic chemist, any proposal has to account for factors such as synthetic accessibility, chemical stability, an interesting pharmacological profile, and, last but not least, the patent situation, as well.

Once a decision has been made to investigate a particular final lead structure in more detail, theoretical activities turn to support the optimization of particular properties of the lead in quantitative terms. Series design methods may guide the first synthetic attempts to get the most information from a well-designed but limited set of structural modifications followed by classical quantitative structure–activity relationship (QSAR) studies (Franke 1984) such as Hansch-type analysis. At this point, the 3D model may serve as a theoretical assay for the essential features which are a neccessary condition for the inherent activity of any structure in the optimization process. If prodrug concepts are to be applied, e.g., in order to achieve oral availability or to eliminate undesired side effects, the biological activities are determined using in vivo experiments. As these data are influenced by pharmacokinetic factors they usually cannot be adequately described by using a model of the putative binding site only.

12.3 Protein Structure-Based Models

If no direct experimental information about the target structure is available, homologous protein structures may provide a good starting point to model the 3D structure. The success of so-called homology modeling approach is, however, critically dependent upon the degree of homology between the systems.

The following example, the modeling of the herbicide binding site of the D1 protein of plants, illustrates the strategy. Many commercial herbicides bind to the D1 protein which is part of the photosystem II of plants and block electron transfer by competitive displacement of quinone Q_B, which serves as a secondary electron acceptor (Trebst 1987).

No direct structural information either from X-ray or NMR studies was available for any of the D1 proteins, which are known to be highly homologous for various plant species. Due to the available biochemical information we decided to develop a model of the D1 protein of spin-

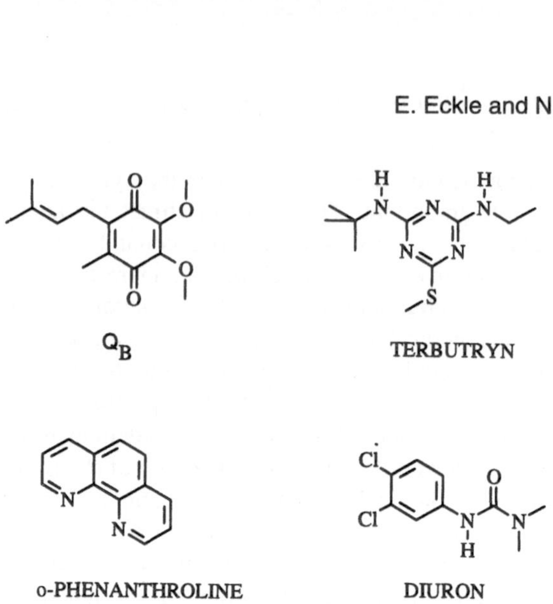

Fig. 4. Chemical structures of compounds whose complexes with the L subunit of the bacterial reaction center have been determined by X-ray crystallography

ach. From biochemical experiments, it appeared likely that its gross topology is similar to that of the L-subunit of the related bacterial reaction center (RC) of *Rhodopseudomonas viridis* (*Rps. viridis*). X-ray structures of the bacterial RC bound to the natural ligand Q_B, and the herbicides terbutryn and *o*-phenanthroline (see Fig. 4) had been solved, respectively (Deisenhofer et al. 1985; Michel et al. 1986a,b). Further information about herbicide binding to the L protein was available from an X-ray structure analysis of diuron bound to a mutant RC of *Rps. viridis* (Sinning et al. 1989). These studies revealed that the active site region is located within an approximately 80-residue-long segment of the L protein. The natural substrate and the inhibitors bind to a shallow groove which is surrounded by the three helices D, DE, and E (see Figs. 5,6).

The corresponding part of the D1 protein containing 17 additional amino acid residues has a low sequence homology to the L segment. Compared with the L region which is 82 residues long, the ratio of identical amino acid residues is only 20%. However, it is worth mentioning that homology studies rely on the assumption that the 3D structures of proteins with comparable functions are conserved even if their sequence homologies are low. With the aid of three conserved func-

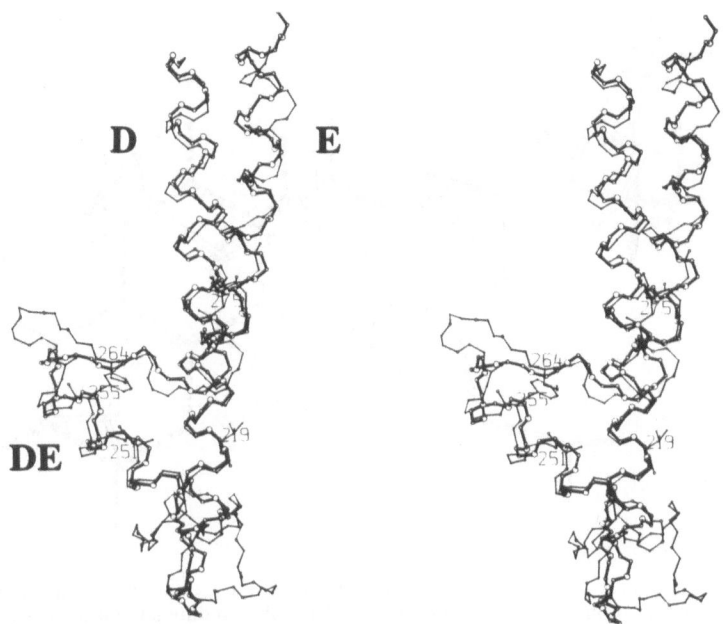

Fig. 5. Superposition of the active site regions of the bacterial L subunit with the modelled D1 protein. The D1 protein is represented by *thin lines*. Helices *D*, *DE*, and *E* are labeled, as well as amino acid residues (*numbers*) which are changed in resistant mutants

tional amino acid residues located in two helices and in the loop linking helices DE and E, a reasonable alignment could be achieved. From this alignment, it appeared that 14 amino acid residues had to be inserted into the "large" loop between helix D and DE and three into the "small" loop following helix DE. As a consequence, the binding site in the plant system was expected to be significantly larger than in the bacterial protein.

Model construction (Egner et al. 1993) started from the X-ray coordinates of the L subunit of *Rps. viridis*. Amino acid residues of the bacterial protein had to be replaced according to the sequence of the D1 protein. The internal geometries of the helices and their relative orientations in space were kept as close as possible to the experimental template. Based on biological data from D1 mutants and labeling experi-

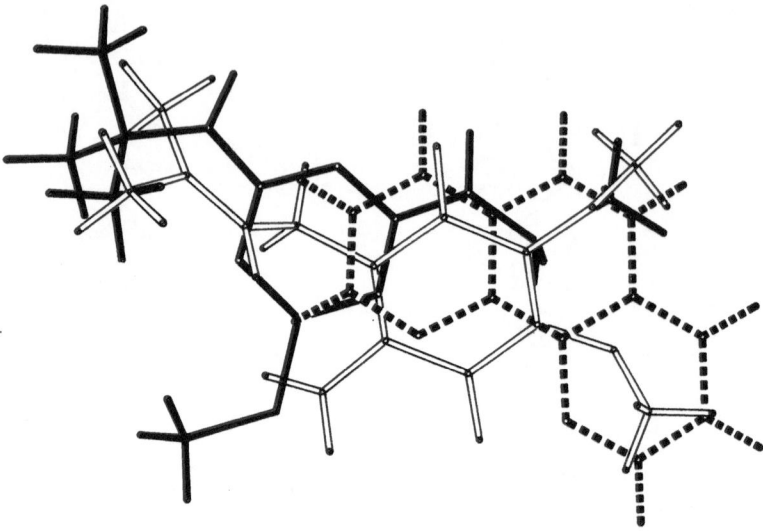

Fig. 6. Relative orientations of terbutryn (*thick bonds*), ubiquinone Q_B (*open bonds*), and *o*-phenanthroline (*broken bonds*) in the binding site on the L sub-unit of *Rps. viridis* as revealed by the crystal structure determinations

ments and geometries obtained from loop searches through the Brookhaven protein data bank, several attempts to model the two loops have been made. In the bacterial system the endogeneous ligand Q_B interacts with the binding site via two hydrogen bonds involving the side chains of the conserved residues Ser-264 (part of the "small" loop) and His-215 (located on helix D). In addition, mutant data suggested that the conserved residue Phe-255 should have retained a similar position as in the bacterial L-unit. However, none of the loop structures derived from the database could simultaneously account for these interactions. Therefore, the "small" loop composed of 10 residues in the plant system had to be corrected manually. With the additional assumption that the D2 protein – being part of the plant RC – would similarly interact with the "small" loop like the corresponding M subunit of the bacterial RC, we finally arrived at a reasonable starting geometry. Likewise, limited experimental information was available to guide the building of the "large" loop, now consisting of 22 residues in the D1 protein. Labeling experiments and proteolytic digestion results gave a preliminary idea about the orientation of some side chains with respect to the binding region and the protein surface. Based on the analysis of loop geometries suggested by the protein data bank the "large" loop was constructed from a combination of u-turn and helix motifs.

The starting model was refined by energy minimization techniques and subsequently thoroughly validated. A set of known PS II inhibitors was selected to prove whether their experimental in vitro inhibition constants, measured in assays of native and mutant systems, could be explained qualitatively. In cases where the binding mode was unequivocally determined by the steric boundaries of the site, the experimental data turned out to be in qualitative agreement with predictions from the model. However, some molecules with rather low inhibitory potency were found to fit well into the putative binding pocket, indicating that some regions in the cavity were poorly defined.

In light of the approximations which have been made to establish the initial model, this finding was not too surprising. To further investigate the boundaries of the site, a class of structures with one well-defined binding mode was chosen. Here, an outer position of the skeleton was substituted by bulky tertiary butyl groups, giving rise to a specific orientation in the binding site. Other parts of the structure pointing towards the less precisely characterized regions of the cavity turned out to be well suited for chemical modification. For steric mapping, specifically designed analog structures have been synthesized. In order to get a good agreement with the biological data distinct parts of the protein structure had to be adjusted. Through these changes to the initial model the active site volume was reduced by about 20%. The refined model allowed the qualitative prediction of the in vitro activities of several inhibitor classes and could successfully be used in the de novo design of a new chemical class of potent inhibitors.

The adjustments to the model were mainly based on the analysis of biological activity data from mutant and wild-type D1 species caused by specifically designed strategic compounds. The refinement, therefore, had to proceed in an indirect way. This is a typical drawback for homology-based models. In contrast, models which are derived directly from an X-ray structure can be validated directly by determination of the crystal structure complexed with selected ligands.

12.4 Perspectives

Any strategy in drug discovery and development will be measured in terms of the time required for success. A rough estimate of the time requirements is given in Fig. 7. By definition, "chemical synthesis and testing" represents the time required to develop a final lead from any initial lead (which may be an endogeneous ligand or structures from literature sources). The synthesis of structures closely related to the final lead is referred to as the "optimization." Apart from the time requirements, another criterion for the choice of a strategy is the degree of innovation: From the chemical point of view, this may be specified in terms of the structural dissimilarity to known structures. How do the various strategies compare in this regard?

As a reference procedure, the "classical" purely chemistry-driven approach is designated as strategy 1. Here, structural variations are performed systematically, thereby increasing the degree of structural dissimilarity by small steps.

In comparison to the traditional strategy, ligand structure-based design methods (strategy 2) are expected to shorten both processes, the development of a final lead of comparable structural dissimilarity and its optimization. The lead finding process can be directed very efficiently once an elaborate theoretical model of the pharmacophore has

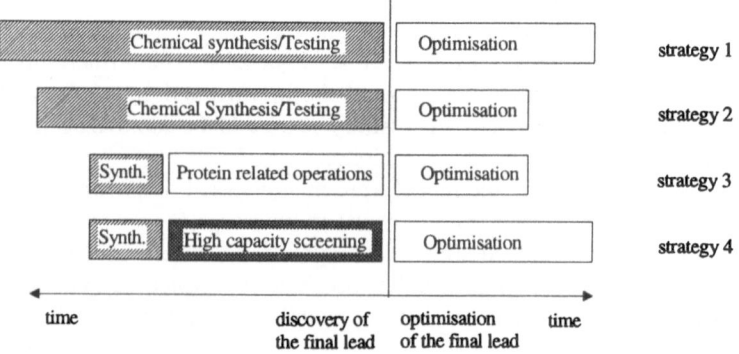

Fig. 7. Schematic representation of time requirements of drug finding strategies

been established. By analogy or chemical intuition, larger steps of chemical modification can be suggested. Within this strategy, 3D database searching provides a powerful tool to generate novel, structurally diverse leads. Compared to strategy 1, the subsequent optimization can take full advantage of the appropriate modifications suggested by the pharmacophore model, thereby retaining the functionality essential for inherent biological activity.

In protein structure-based strategies (strategy 3), the protein-related operations such as overexpression, purification, crystallization, and structure elucidation represent the major time-consuming part of the lead finding process. It is hard to say in general terms how these operations will compare with the corresponding steps of lead discovery in the chemically driven approaches. Once suitable crystals have been obtained, the initial crystal structure determination is much more time-consuming than subsequent crystallographic determinations of complexes with newly designed ligands. It is expected that the time requirements for subsequent synthesis will be comparably small: The suggestions for structures from de novo methods are fully based on the 3D structure of the binding cavity. From the various proposals a well-selected set of structures will be synthesized which combine both, for instance, chemical feasibility and sufficient structural dissimilarity to known ligand structures. Similar to the ligand-based approach, the subsequent optimization can largely benefit from a detailed knowledge about the essential features for tight binding. Without any doubt, this approach has the encouraging potential to come up with highly innovative classes of chemical structures within a reduced period of time.

Following the fourth strategy, high-capacity screening of large (combinatorial) libraries will provide initial lead structures which do not have any chemical history and, therefore, might have a relatively high degree of structural dissimilarity to known ligand structures as well. The time necessary to develop the final lead structure from a first hit should be comparable to strategy 3. The subsequent optimization, however, neither benefits from any elaborate structure–activity relationships nor has any advantage from information about the binding cavity. This step, therefore, is expected to take more time compared to the strategies 2 and 3, respectively.

The rate-determining steps of the approaches outlined above are different: For strategy 2, it is the generation and validation of the pharmacophore model which is dependent upon the synthetical capacity provided for strategic structures. In strategy 3, as already mentioned, the protein-related operations, particularly the crystallization of the first protein–ligand complex, determines the rate of progress. In the random approach, we expect the process of chemical optimization to be rate-determining provided some hits are found in the high-capacity screening at all. In a medium-term perspective, the whole spectrum of approaches will be applied in parallel as there is no clear-cut evidence as to which approach will be the most successful in finding a lead and developing a drug.

References

Alper J (1994) Drug discovery on the assembly line. Science 264: 1399–1401

Armstrong RA, Jones RL, MacDermot J, Wilson NH (1986) Prostaglandin endoperoxide analogues which are both thromboxane receptor agonists and prostacyclin mimetics. Br J Pharmacol 87: 543–551

Deisenhofer J, Epp O, Miki K, Huber R, Michel H (1985) Structure of the protein subunits in the photosynthetic reaction centre of Rhodopseudomonas viridis at 3 Å resolution. Nature 318: 618–624

Egner U, Hoyer G-A, Saenger W (1993) Modeling and energy minimization studies on the herbicide binding protein (D1) in photosystem II of plants. Biochim Biophys Acta 1142: 106–114

Franke R (1984) Theoretical drug design methods. Elsevier, Amsterdam

Gallop MA, Barrett RW, Dower WJ, Fodor SPA, Gordon EM (1994) Applications of combinatorial technologies to drug discovery 1. Background and peptide combinatorial libraries. J Med Chem 37:1233–1251

Gunsteren WF van, Berendsen HJC (1990) Computer simulation of molecular dynamics: methodology, applications, and perspectives in chemistry. Angew Chem Int Ed Engl 29:992–1023

Gunsteren WF van, Mark AE (1992) On the interpretation of biochemical data by molecular dynamics computer simulation. Eur J Biochem 204: 947–961

Gryglewski RJ, Stock G (eds) (1987) Prostacyclin and its stable analogue iloprost. Springer, Berlin Heidelberg New York

Johnson MA, Maggiora GM (eds) (1990) Concepts and applications of molecular similarity. Wiley, New York

Lybrand TP (1990) Computer simulation of biomolecular systems using molecular dynamics and free energy perturbation methods. In: Lipkowitz KB, Boyd DB (eds) Reviews in computational chemistry. VCH, New York, pp 295–320

Marshall GR, Barry CD, Bosshard HE, Dammkoehler RA, Dunn DA (1979) The conformational parameter in drug design: the active analog approach. In: Olson EC, Christoffersen RE (eds) Computer-assisted drug design ACS symposium series 112. Am Chem Soc, Washington DC, pp 205–226

Martin YC (1992) 3D database searching in drug design. J Am Chem Soc 35: 2146–2154

Merritt JE, Hallam TJ, Brown AM, Boyfield I, Cooper DG, Hickey DMB, Jaxa-Chamiec AA, Kaumann AJ, Keen M, Kelly E, Kozlowski U, Lynham JA, Moores KE, Murray KJ, MacDermot J, Rink TJ (1991) Octimibate, a potent non-prostanoid inhibitor of platelet aggregation, acts via the prostacyclin receptor. Br J Pharmacol 102:251–259

Michel H, Epp O, Deisenhofer J (1986a) Pigment-protein interactions in the photosynthetic reaction center from Rhodopseudomonas viridis. EMBO J 5: 2445–2451

Michel H, Weyer KA, Gruenberg H, Dunger I, Oesterhelt D, Lottspeich F (1986b) The 'light' and 'medium' subunits of the photosynthetic reaction center from Rhodopseudomonas viridis: isolation of the genes, nucleotide and amino acid sequence. EMBO J 5: 1149–1158

Primas H, Müller-Herold U (1984) Elementare Quantenchemie Teubner, Stuttgart, p 289

Seiler S, Brassard CL, Arnold AJ, Meanwell NA, Fleming JS, Keely SL (1990) Octimibate inhibition of platelet aggregation: stimulation of adenylate cyclase through prostacyclin receptor activation. J Pharmacol Exp Ther 255: 1021–1026

Sheridan RP, Nilakantan R, Dixon JS, Venkataraghavan R (1986) The ensemble approach to distance geometry: application to the nicotinic pharmacophore. J Med Chem 29: 899–906

Sinning I, Michel H, Mathis P, Rutherford AW (1989) Terbutryn resistance in a purple bacterium can induce sensitivity toward the plant herbicide DCMU. FEBS Lett 256:192–194

Skuballa W, Schäfer M (1989) Prostacyclin-Derivate. Nachr Chem Techn Lab 37:584–590

Trebst A (1987) The three-dimensional structure of the herbicide binding niche on the reaction center polypeptides of photosystem II Z. Naturforsch 42c:742–750

Verlinde CLMJ, Hol WGJ (1994) Structure-based drug design: progress, results and challenges. Structure 2: 577–587

13 Computer-Aided Drug Design in Industry: A Summary of Perspectives

R. Franke and E.C. Herrmann

Based on the discussions and, in particular, a general discussion held at the end of the workshop this final chapter summarizes questions, views, and conclusions relating to the following topics:

1. Effectiveness/impact of computer-aided drug design (CADD) in industrial drug discovery projects
2. Contributions of CADD groups to drug discovery
3. Conditions supporting/prohibiting useful contributions from CADD groups in industry
4. The role of CADD in combination with combinatorial chemistry and high throughput screening

There was general agreement that CADD has developed into an indispensable tool (or, rather, toolbox) in modern drug research or in the search for new agrochemical agents which is well integrated in drug discovery chemistry. It seems, however, that it is very difficult, if not impossible, to find objective or quantitative measures of the practical impact of CADD. The main reason is that drug research is a multidisciplinary field and that the success of a drug discovery project depends on a variety of factors, among which CADD is only one; the individual contributions of these factors cannot be separated straightforwardly in real life situations.

Sometimes the question is asked as to which compounds on the market have been found through CADD. This question is beside the point not only because of the multidisciplinary nature of drug research but also because CADD only operates before clinical trials are begun; in addition, projects may be stopped by management decisions even if promising compounds have been found. All the speakers presented positive examples from their own work where CADD had made essential contributions to the discovery of biologically active compounds. It was felt, in addition, that there is a tendency for projects to get started and terminated faster which means, in other words, that a greater number of diverse compounds are considered in shorter times.

The best and most important criterion for the effectiveness of a CADD group in a given company was considered to be the acceptance by synthetic chemists and pharmacologists or biologists.

The cost ratio of a theoretical (CADD) chemist compared with a synthetic chemist varies from company to company, depending, among other factors, on how costs are calculated. Ratios from 0.8/1 up to 2/1 were mentioned. A CADD group should have a "critical mass," optimally close to one theorist for about 10 synthetic chemists. It was also pointed out that adding an additional member to an already existing CADD group is not so expensive as all the necessary hardware and software is already available.

A number of criteria were mentioned regarding CADD's effectiveness. It was stressed that a CADD group must be able to use the whole "tool box" of methods in order to be able to successfully cope with the many different tasks, types of biological data, and chemical structures which appear in practice.

Such methods include approaches based on molecular modeling and structure as well as chemometric techniques. The use of various approaches simultaneously was advocated whenever this is feasible, being aware of the fact that the importance of the different types of analysis may vary through the course of a project. The results will supplement and support each other. Sometimes different approaches will provide different answers. This offers the possibility of generating new ideas and reinforcing (or falsifying) conclusions by well-directed experiments in an early stage. Very close collaboration with synthetic chemists and pharmacologists/biologists based on frequent and free discussions was seen as being the most important factor. All disciplines involved in the drug discovery process share the responsibility, which helps to select and justify certain directions of work. This collaboration must be backed by the upper management. CADD is not a technique to provide dramatic, overnight discoveries of new therapeutic agents but rather a means to generate ideas (which frequently are put forward in terms of chemical structure) with the objective of optimizing the drug discovery process. These ideas have to be discussed with the synthetic chemist, and based on these discussions and proposals from the CADD group it is the synthetic chemist who makes the final decisions as to which compounds are to be synthesized.

This implies, of course, that CADD results are taken seriously and are sufficiently well reflected in these decisions. Without actual experiments and feedback CADD would be meaningless. Another important point is that synthetic chemists must be willing to make compounds of expectedly low potency if this is necessary to prove or disprove a theoretical model on which further decision making is to be based. CADD compounds can be pretty difficult to synthesize; thus, chemists should not be judged by the number of compounds they synthesize but rather by the overall success of a project.

Discussion among those involved in CADD, synthetic chemists, and pharmacologists requires a mutual understanding and some kind of a common language. This, in turn, requires a willingness to learn on the part of everyone involved, which can best be seen as continuous mutual education. Sitting together in front of a computer screen to discuss models in detail was felt to be particularly useful. A common language is also required to assign CADD groups to relevant projects. This should be done at as early a stage as possible, with a clear definition of the role the CADD group is expected to play. Work in interdisciplinary project groups involving CADD scientists was considered to be a very effective approach. At the same time, the theoretical chemists should be organized in a central entity to reinforce special expertise. One of the most important contributions of CADD is that it has introduced and continues to refine new ways of thinking in medicinal chemistry.

A critical evaluation of CADD results was regarded to be very important. Otherwise, such results may be misleading. One should not always believe what is shown in a nice picture or what a high statistical fit of a chemometric model seems to suggest. Also, X-ray structure findings may contain a good deal of fiction. Overestimation of what CADD really can do for a given project may be very harmful.

CADD is involved in all stages of the drug discovery process: lead generation, lead optimization, and the selection of samples for synthesis and testing (series design). For lead generation, the importance of closely combining crystallography with molecular biology (structure-based design) was emphasized. Lead structures should be looked for at all stages of a project, including clinical testing, which may create new ideas and contribute to protect patents and/or to arrive at back-up leads. CADD groups can help to organize project-specific data bases, which were regarded as very useful tools to be provided with every project. Such data bases must always be up to date, accessible to everybody involved in the project, and extremely simple to operate. Another important aspect of CADD is that it can help to decide when to stop a project by showing, for example, that more active compounds are not likely to exist or that it will not be possible to separate a desired from an undesired effect.

It was stressed that the objective of drug discovery is to find drugs, not just ligands for an enzyme or receptor. For this reason it is necessary that questions such as bioavailability are addressed early enough in a project in order not to end up with a compound of fantastic affinity for some target which is not absorbed in vivo. Physicochemical parameters in connection with chemometric methods can be used for this. Another important issue is toxic or other unwanted effects. Even though it was agreed that there is no general scheme to reliably predict toxicity from chemical structure, this is still possible within subgroups of compounds acting all by the same mechanism of action. In addition, with certain substructural approaches CADD also operates as a warning technique, i.e., if certain structures occur suspicion is aroused.

Another aspect of CADD is to provide activity–activity relationships; this was regarded as crucial, but has been largely neglected in the past. There are frequently discrepancies between in vitro and in vivo results and there may be many reasons for this. If pharmacokinetic processes are involved a correction via physicochemical properties using chemometric methods is an interesting approach. On the other hand, detecting such discrepancies can help to form new hypotheses to head in the direction one really wants to go. Obviously, it would not make much sense to optimize a lead on the basis of in vitro results if there is no correlation with in vivo measurements and no explanation for this can be found. However, such correlations need not be exactly quantitative in order to accept in vitro data as a guide.

A special case of activity–activity relationship are structure–selectivity relationships which can be used to optimize selectivity either by modeling or chemometric techniques. A warning was given to be very careful with selectivity indices derived from measurements of desired and undesired effects as, in such indices, the errors of measurements may add in an unpredictable manner.

Combinatorial chemistry and high throughput screening are in the process of becoming very important in medicinal chemistry, primarily aiming at lead discovery. What the role of CADD would be in combination with this development was a question of great interest. It was generally agreed that CADD has definite tasks to perform in this connection, again offering the possibility to optimize research in close interaction with the experiment.

Chemical libraries contain very large numbers of compounds which have certain basic features in common and thus become redundant. As there are costs involved in investigating compounds it is an important objective to reduce such redundancies (the number of compounds to be investigated). This can be done by CADD using set selection methods. If, for example, a library primarily contains few substructures with a large variation of substituents, methods based on principal component analysis and/or factorial design can be used. Another possibility is to select compounds from clusters derived from certain similarity measures, including relatively rough substructural considerations. Finally, the possibility to apply "electronic screening" based on docking studies was also mentioned, pointing out at the same time that problems will arise with flexible ligands and that the docking procedures need to be improved further. In summary, CADD methods to design chemical libraries or to select nonredundant series of compounds were regarded to be of great practical importance. This is also true if compounds from an existing data base are to be submitted to screening programs. Care must be taken, however, not to become too restrictive as small structural changes can always lead to drastic changes in a biological profile. The real point is to find the right balance between compound selection by CADD and further technological developments in automatization, allowing for increasing numbers of compounds to be investigated more cost effectively. With this development lead optimization by chemometric methods will gain in importance.

High throughput screening will result in a huge amount of biological data. Such data should not only be stored in data bases. It was stressed that there will be structures in the data (multidimensional activity–activity relationships) which should be elucidated by applying appropriate methods such as multivariate statistics, neural nets, or fuzzy logics in order to extract as much information as possible. This may aid in guessing at biological profiles from batteries of simple in vitro or ex vivo tests or to decide which leads are to be followed up if several are found. In addition, CADD can be used as a warning technique by considering physicochemical properties or results of special tests, taking also into account problems of bioavailability. It was, therefore, emphasized that "screening for physicochemical properties" (e.g., determination of log P for selected compounds plus calculation of log P for the rest of the compounds) should become part of high throughput screening systems as well as a number of tests indicative of metabolic properties or potential hazards (e.g., liver microsomes, P 450, cell toxicity, or cell permeation studies etc.). It was then concluded that combinatorial chemistry and/or high throughput screening can become much more efficient when supported by CADD strategies than the pure brute-force approach.

In summary, all participants felt that computer-aided drug design has a substantial impact on drug discovery chemistry in industrial practice, provided that it is state of the art and well integrated.

Subject Index

Ernst Schering Research Foundation Workshop

Editors: Günter Stock
Ursula-F. Habenicht